An Introduction to the

Analysis of Algorithms

2nd Edition

An Introduction to the Analysis of Algorithms

2nd Edition

Michael Soltys
McMaster University, Canada

NEW JERSEY • LONDON • SINGAPORE • BEIJING • SHANGHAI • HONG KONG • TAIPEI • CHENNAI

Published by

World Scientific Publishing Co. Pte. Ltd.

5 Toh Tuck Link, Singapore 596224

USA office: 27 Warren Street, Suite 401-402, Hackensack, NJ 07601

UK office: 57 Shelton Street, Covent Garden, London WC2H 9HE

British Library Cataloguing-in-Publication Data
A catalogue record for this book is available from the British Library.

AN INTRODUCTION TO THE ANALYSIS OF ALGORITHMS
2nd Edition

ISBN-13 978-981-4401-15-9
ISBN-10 981-4401-15-3

Printed in Singapore by World Scientific Printers.

To my family

Preface

This book is an introduction to the analysis of algorithms, from the point of view of proving algorithm correctness. Our theme is the following: how do we argue mathematically, without a burden of excessive formalism, that a given algorithm does what it is supposed to do? And why is this important? In the words of C.A.R. Hoare:

> As far as the fundamental science is concerned, we still certainly do not know how to prove programs correct. We need a lot of steady progress in this area, which one can foresee, and a lot of breakthroughs where people suddenly find there's a simple way to do something that everybody hitherto has thought to be far too difficult[1].

Software engineers know many examples of things going terribly wrong because of program errors; their particular favorites are the following two[2]. The blackout in the American North-East during the summer of 2003 was due to a software bug in an energy management system; an alarm that should have been triggered never went off, leading to a chain of events that climaxed in a cascading blackout. The Ariane 5, flight 501, the maiden flight of the rocket in June 4, 1996, ended with an explosion 40 seconds into the flight; this $500 million loss was caused by an overflow in the conversion from a 64-bit floating point number to a 16-bit signed integer.

While the goal of absolute certainty in program correctness is elusive, we can develop methods and techniques for reducing errors. The aim of this book is modest: we want to present an introduction to the analysis of algorithms—the "ideas" behind programs, and show how to prove their correctness.

---

[1]From *An Interview with C.A.R. Hoare*, in [Shustek (2009)].

[2]These two examples come from [van Vliet (2000)], where many more instances of spectacular failures may be found.

The algorithm may be correct, but the implementation itself might be flawed. Some syntactical errors in the program implementation may be uncovered by a compiler or translator—which in turn could also be buggy—but there might be other hidden errors. The hardware itself might be faulty; the libraries on which the program relies at run time might be unreliable, etc. It is the task of the software engineer to write code that works given such a delicate environment, prone to errors. Finally, the algorithmic content of a piece of software might be very small; the majority of the lines of code could be the "menial" task of interface programming. Thus, the ability to argue correctly about the soundness of an algorithm is only one of many facets of the task at hand, yet an important one, if only for the pedagogical reason of learning to argue rigorously about algorithms.

We begin this book with a chapter of preliminaries, containing the key ideas of induction and invariance, and the framework of pre/post-conditions and loop invariants. We also prove the correctness of some classical algorithms, such as the integer division algorithm, and Euclid's procedure for computing the greatest common divisor of two numbers.

We present three standard algorithm design techniques in eponymous chapters: greedy algorithms, dynamic programming and the divide and conquer paradigm. We are concerned with correctness of algorithms, rather than, say, efficiency or the underlying data structures. For example, in the chapter on the greedy paradigm we explore in depth the idea of a *promising partial solution*, a powerful technique for proving the correctness of greedy algorithms. We include online algorithms and *competitive analysis*, as well as randomized algorithms with a section on *cryptography*.

Algorithms solve problems, and many of the problems in this book fall under the category of *optimization problems*, whether cost minimization, such as Kruskal's algorithm for computing minimum cost spanning trees—section 2.1, or profit maximization, such as selecting the most profitable subset of activities—section 4.4.

The book is sprinkled with problems; most test the understanding of the material, but many include coding in the Python programming language. The reader is expected to learn Python on their own (a great source to do so is [Downey (2008)]—the PDF can be downloaded for free from the web). One of the advantages of this programming language is that it is easy to start writing small snippets of code that work—and most of the coding in this book falls into the "small snippet" category. The solutions to most problems are included in the "Answers to selected problems" at the end of each chapter. The solutions to most of the programming exercises will be available for download from the author's web page (a quick Google search will reveal the URL).

The intended audience for this book consists of undergraduate students in computer science and mathematics. The book is very self-contained: the first chapter, Preliminaries, reviews induction and the invariance principle. It also introduces the aforementioned ideas of pre/post-conditions, loop invariants and termination—in other words, it sets the mathematical stage for the rest of the book. Not much mathematics is assumed (besides some tame forays into linear algebra and number theory), but a certain penchant for discrete mathematics is considered helpful.

This book draws on many sources. First of all, [Cormen *et al.* (2009)] is a fantastic reference for anyone who is learning algorithms. I have also used as reference the elegantly written [Kleinberg and Tardos (2006)]. A classic in the field is [Knuth (1997)], and I base my presentation of online algorithms on the material in [Borodin and El-Yaniv (1998)]. I have learned greedy algorithms, dynamic programming and logic from Stephen A. Cook at the University of Toronto. Appendix B, a digest of relations, is based on hand-written lecture slides of Ryszard Janicki.

No book on algorithms is complete without a short introduction to the "big-Oh" notation. Consider functions from $\mathbb{N}$ to $\mathbb{N}$. We say that $g(n) \in O(f(n))$ if there exist constants c, n_0 such that for all $n \geq n_0$, $g(n) \leq cf(n)$, and the *little-oh* notation, $g(n) \in o(f(n))$, which denotes that $\lim_{n\to\infty} g(n)/f(n) = 0$. We also say that $g(n) \in \Omega(f(n))$ if there exist constants c, n_0 such that for all $n \geq n_0$, $g(n) \geq cf(n)$. Finally, we say that $g(n) \in \Theta(f(n))$ if it is the case that $g(n) \in O(f(n)) \cap \Omega(f(n))$.

The ubiquitous *floor* and *ceil* functions are defined, respectively, as follows: $\lfloor x \rfloor = \max\{n \in \mathbb{Z} | n \leq x\}$ and $\lceil x \rceil = \min\{n \in \mathbb{Z} | n \geq x\}$. Finally, $\lfloor x \rceil$ refers to the "rounding-up" of x, and it is defined as $\lfloor x \rceil = \lfloor x + \frac{1}{2} \rfloor$.

We have the usual Boolean connectives: $\wedge$ is "and," $\vee$ is "or" and $\neg$ is "not." We also use $\rightarrow$ as Boolean implication, i.e., $x \rightarrow y$ is logically equivalent to $\neg x \vee y$, and $\leftrightarrow$ is Boolean equivalence, and $\alpha \leftrightarrow \beta$ expresses $((\alpha \rightarrow \beta) \wedge (\beta \rightarrow \alpha))$

We use "$\Rightarrow$" to abbreviate the word "implies," i.e., $2|x \Rightarrow x$ is even, while "$\not\Rightarrow$" abbreviates "does not imply."

Contents

Chapter 1

Preliminaries

1.1 Induction

Let $\mathbb{N} = \{0, 1, 2, \ldots\}$ be the set of natural numbers. Suppose that S is a subset of $\mathbb{N}$ with the following two properties: first $0 \in S$, and second, whenever $n \in S$, then $n + 1 \in S$ as well. Then, invoking the *Induction Principle (IP)* we can conclude that $S = \mathbb{N}$.

We shall use the IP with a more convenient notation; let P be a property of natural numbers, in other words, P is a unary relation such that $\mathrm{P}(i)$ is either true or false. The relation P may be identified with a set S_{P} in the obvious way, i.e., $i \in S_{\mathrm{P}}$ iff $\mathrm{P}(i)$ is true. For example, if P is the property of being prime, then $\mathrm{P}(2)$ and $\mathrm{P}(3)$ are true, but $\mathrm{P}(6)$ is false, and $S_{\mathrm{P}} = \{2, 3, 5, 7, 11, \ldots\}$. Using this notation the IP may be stated as:

$$[\mathrm{P}(0) \wedge \forall n(\mathrm{P}(n) \rightarrow \mathrm{P}(n+1))] \rightarrow \forall m \mathrm{P}(m), \tag{1.1}$$

for any (unary) relation P over $\mathbb{N}$. In practice, we use (1.1) as follows: first we prove that $\mathrm{P}(0)$ holds (this is the *basis case*). Then we show that $\forall n(\mathrm{P}(n) \rightarrow \mathrm{P}(n+1))$ (this is the *induction step*). Finally, using (1.1) and *modus ponens*, we conclude that $\forall m \mathrm{P}(m)$.

As an example, let P be the assertion "the sum of the first i odd numbers equals i^2." We follow the convention that the sum of an empty set of numbers is zero; thus $\mathrm{P}(0)$ holds as the set of the first zero odd numbers is an empty set. $\mathrm{P}(1)$ is true as $1 = 1^2$, and $\mathrm{P}(3)$ is also true as $1+3+5 = 9 = 3^2$. We want to show that in fact $\forall m \mathrm{P}(m)$ i.e., P is always true, and so $S_{\mathrm{P}} = \mathbb{N}$.

Notice that $S_{\mathrm{P}} = \mathbb{N}$ does not mean that all numbers are odd—an obviously false assertion. We are using the natural numbers to *index* odd numbers, i.e., $o_1 = 1, o_2 = 3, o_3 = 5, o_4 = 7, \ldots$, and our induction is over this indexing (where o_i is the i-th odd number, i.e., $o_i = 2i - 1$). That is, we are proving that for all $i \in \mathbb{N}$, $o_1 + o_2 + o_3 + \cdots + o_i = i^2$; our assertion

P(i) is precisely the statement "$o_1 + o_2 + o_3 + \cdots + o_i = i^2$."

We now use induction: the basis case is P(0) and we already showed that it holds. Suppose now that the assertion holds for n, i.e., the sum of the first n odd numbers is n^2, i.e., $1 + 3 + 5 + \cdots + (2n - 1) = n^2$ (this is our *inductive hypothesis* or *inductive assumption*). Consider the sum of the first $(n + 1)$ odd numbers,

$$\boxed{1 + 3 + 5 + \cdots + (2n - 1)} + (2n + 1) = \boxed{n^2} + (2n + 1) = (n + 1)^2,$$

and so we just proved the induction step, and by IP we have $\forall m \mathrm{P}(m)$.

Problem 1.1. *Prove that* $1 + \sum_{j=0}^{i} 2^j = 2^{i+1}$.

Sometimes it is convenient to start our induction higher than at 0. We have the following generalized induction principle:

$$[\mathrm{P}(k) \wedge (\forall n \geq k)(\mathrm{P}(n) \rightarrow \mathrm{P}(n + 1))] \rightarrow (\forall m \geq k)\mathrm{P}(m), \tag{1.2}$$

for any predicate P and any number k. Note that (1.2) follows easily from (1.1) if we simply let P$'(i)$ be P$(i + k)$, and do the usual induction on the predicate P$'(i)$.

Problem 1.2. *Use induction to prove that for* $n \geq 1$,

$$1^3 + 2^3 + 3^3 + \cdots + n^3 = (1 + 2 + 3 + \cdots + n)^2.$$

Problem 1.3. *For every* $n \geq 1$, *consider a square of size* $2^n \times 2^n$ *where one square is missing. Show that the resulting square can be filled with "L" shapes—that is, with clusters of three squares, where the three squares do not form a line.*

Problem 1.4. *Suppose that we restate the generalized IP (1.2) as*

$$[\mathrm{P}(k) \wedge \forall n(\mathrm{P}(n) \rightarrow \mathrm{P}(n + 1))] \rightarrow (\forall m \geq k)\mathrm{P}(m). \tag{1.2'}$$

What is the relationship between (1.2) and (1.2')?

Problem 1.5. *The* Fibonacci sequence *is defined as follows:* $f_0 = 0$ *and* $f_1 = 1$ *and* $f_{i+2} = f_{i+1} + f_i$, $i \geq 0$. *Prove that for all* $n \geq 1$ *we have:*

$$\begin{pmatrix} 1 & 1 \\ 1 & 0 \end{pmatrix}^n = \begin{pmatrix} f_{n+1} & f_n \\ f_n & f_{n-1} \end{pmatrix},$$

where the left-hand side is the n*-th power of a* 2×2 *matrix.*

Problem 1.6. *Write a Python program that computes the* n*-th Fibonacci number using the matrix multiplication trick of problem 1.5.*

Problem 1.7. *Prove the following: if m divides n, then f_m divides f_n, i.e., $m|n \Rightarrow f_m|f_n$.*

The *Complete Induction Principle*(CIP) is just like IP except that in the induction step we show that if $\mathrm{P}(i)$ holds for all $i \leq n$, then $\mathrm{P}(n+1)$ also holds, i.e., the induction step is now $\forall n((\forall i \leq n)\mathrm{P}(i) \rightarrow \mathrm{P}(n+1))$.

Problem 1.8. *Use the CIP to prove that every number (in $\mathbb{N}$) greater than 1 may be written as a product of one or more prime numbers.*

Problem 1.9. *Suppose that we have a (Swiss) chocolate bar consisting of a number of squares arranged in a rectangular pattern. Our task is to split the bar into small squares (always breaking along the lines between the squares) with a minimum number of breaks. How many breaks will it take? Make an educated guess, and prove it by induction.*

The *Least Number Principle (LNP)* says that every non-empty subset of the natural numbers must have a least element. A direct consequence of the LNP is that every decreasing non-negative sequence of integers must terminate; that is, if $R = \{r_1, r_2, r_3, \ldots\} \subseteq \mathbb{N}$ where $r_i > r_{i+1}$ for all i, then R is a *finite* subset of $\mathbb{N}$. We are going to be using the LNP to show termination of algorithms.

Problem 1.10. *Show that IP, CIP, and LNP are equivalent principles.*

There are three standard ways to list the nodes of a binary tree. We present them below, together with a recursive procedure that lists the nodes according to each scheme.

Infix: left sub-tree, root, right sub-tree.

Prefix: root, left sub-tree, right sub-tree.

Postfix: left sub-tree, right sub-tree, root.

See the example in figure 1.1.

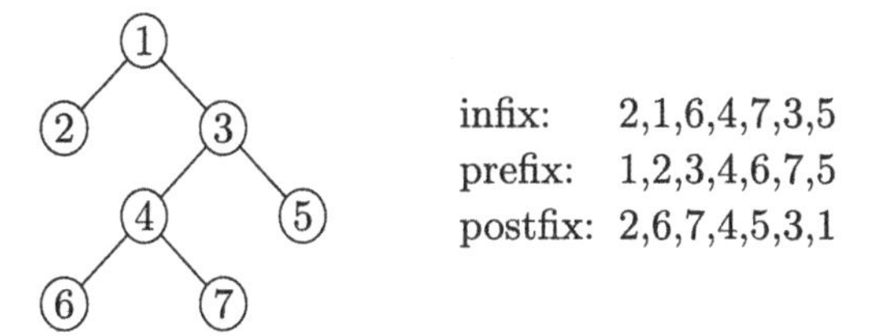

Fig. 1.1 A binary tree with the corresponding representations.

Note that some authors use a different name for infix, prefix, and postfix; they call it inorder, preorder, and postorder, respectively.

Problem 1.11. *Show that given any two representations we can obtain from them the third one, or, put another way, from any two representations we can reconstruct the tree. Show, using induction, that your reconstruction is correct. Then show that having just one representation is not enough.*

Problem 1.12. *Write a Python program that takes as input two of the three descriptions, and outputs the third. One way to present the input is as a text file, consisting of two rows, for example*

```
infix: 2,1,6,4,7,3,5
postfix: 2,6,7,4,5,3,1
```

and the corresponding output would be: `prefix: 1,2,3,4,6,7,5`. *Note that each row of the input has to specify the "scheme" of the description.*

1.2 Invariance

The *Invariance Technique (IT)* is a method for proving assertions about the outcomes of procedures. The IT identifies some property that remains true throughout the execution of a procedure. Then, once the procedure terminates, we use this property to prove assertions about the output.

As an example, consider an 8×8 board from which two squares from opposing corners have been removed (see figure 1.2). The area of the board is $64-2=62$ squares. Now suppose that we have 31 dominoes of size 1×2. We want to show that the board *cannot* be covered by them.

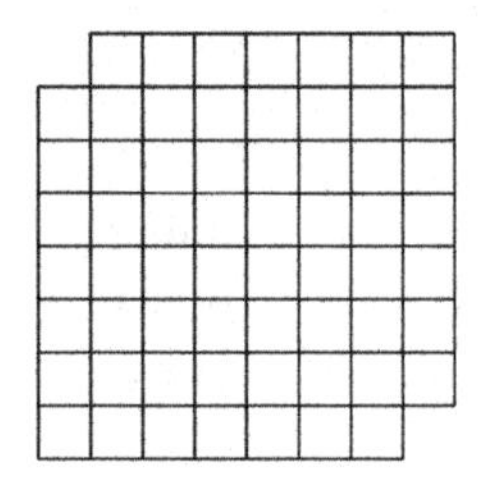

Fig. 1.2 An 8×8 board.

Verifying this by *brute force* (that is, examining all possible coverings) is an extremely laborious job. However, using IT we argue as follows: color the squares as a chess board. Each domino, covering two adjacent squares, covers 1 white and 1 black square, and, hence, each placement covers as many white squares as it covers black squares. Note that the number of white squares and the number of black squares differ by 2—opposite corners lying on the same diagonal have the same color—and, hence, no placement of dominoes yields a cover; done!

More formally, we place the dominoes one by one on the board, any way we want. The invariant is that after placing each new domino, the number of covered white squares is the same as the number of covered black squares. We prove that this *is* an invariant by induction on the number of placed dominoes. The basis case is when zero dominoes have been placed (so zero black and zero white squares are covered). In the induction step, we add one more domino which, no matter how we place it, covers one white and one black square, thus maintaining the property. At the end, when we are done placing dominoes, we would have to have as many white squares as black squares covered, which is not possible due to the nature of the coloring of the board (i.e., the number of black and whites squares is not the same). Note that this argument extends easily to the $n \times n$ board.

Problem 1.13. *Let n be an odd number, and suppose that we have the set $\{1, 2, \ldots, 2n\}$. We pick any two numbers a, b in the set, delete them from the set, and replace them with $|a-b|$. Continue repeating this until just one number remains in the set; show that this remaining number must be odd.*

The next three problems have the common theme of social gatherings. We always assume that relations of likes and dislikes, of being an enemy or a friend, are reflexive relations: that is, if a likes b, then b also likes a, etc. See appendix B for background on relations—reflexive relations are defined on page 157.

Problem 1.14. *At a country club, each member dislikes at most three other members. There are two tennis courts; show that each member can be assigned to one of the two courts in such a way that at most one person they dislike is also playing on the same court.*

We use the vocabulary of "country clubs" and "tennis courts," but it is clear that Problem 1.14 is a typical situation that one might encounter in computer science: for example, a multi-threaded program which is run on two processors, where a pair of threads are taken to be "enemies" when they use many of the same resources. Threads that require the same resources ought to be scheduled on different processors—as much as possible. In a sense, these seemingly innocent problems are parables of computer science.

Problem 1.15. *You are hosting a dinner party where $2n$ people are going to be sitting at a round table. As it happens in any social clique, animosities are rife, but you know that everyone sitting at the table dislikes at most $(n-1)$ people; show that you can make sitting arrangements so that nobody sits next to someone they dislike.*

Problem 1.16. *Handshakes are exchanged at a meeting. We call a person an* odd person *if he has exchanged an odd number of handshakes. Show that, at any moment, there is an even number of odd persons.*

1.3 Correctness of algorithms

How can we prove that an algorithm is correct[1]? We make two assertions, called the *pre-condition* and the *post-condition*; by correctness we mean that whenever the pre-condition holds *before* the algorithm executes, the post-condition will hold *after* it executes. By *termination* we mean that whenever the pre-condition holds, the algorithm will stop running after finitely many steps. Correctness without termination is called *partial correctness*, and *correctness* per se is partial correctness *with* termination.

These concepts can be made more precise: let A be an algorithm, and let $\mathcal{I}_A$ be the set of all *well-formed* inputs for A; the idea is that if $I \in \mathcal{I}_A$ then it "makes sense" to give I as an input to A. The concept of a "well-formed" input can also be made precise, but it is enough to rely on our intuitive understanding—for example, an algorithm that takes a pair of integers as input will not be "fed" a matrix. Let $O = A(I)$ be the output of A on I, if it exists. Let α_A be a pre-condition and β_A a post-condition of A; if I satisfies the pre-condition we write $\alpha_A(I)$ and if O satisfies the post-condition we write $\beta_A(O)$. Then, partial correctness of A with respect to pre-condition α_A and post-condition β_A can be stated as:

$$(\forall I \in \mathcal{I}_A)[(\alpha_A(I) \wedge \exists O(O = A(I))) \rightarrow \beta_A(A(I))], \tag{1.3}$$

[1] A wonderful introduction to this topic can be found in [Harel (1987)], in chapter 5, "The correctness of algorithms, or getting it done right."

which in words states the following: for any well formed input I, if I satisfies the pre-condition and $A(I)$ produces an output, i.e., terminates, which is stated as $\exists O(O = A(I))$, then this output satisfies the post-condition.

Full correctness is (1.3) together with the assertion that for all $I \in \mathcal{I}_A$, $A(I)$ terminates (and hence there exists an O such that $O = A(I)$).

A fundamental notion in the analysis of algorithms is that of a *loop invariant*; it is an assertion that stays true after each execution of a "while" (or "for") loop. Coming up with the right assertion, and proving it, is a creative endeavor. If the algorithm terminates, the loop invariant is an assertion that helps to prove the implication $\alpha_A(I) \rightarrow \beta_A(A(I))$.

Once the loop invariant has been shown to hold, it is used for proving partial correctness of the algorithm. So the criterion for selecting a loop invariant is that it helps in proving the post-condition. In general many different loop invariants (and for that matter pre and post-conditions) may yield a desirable proof of correctness; the art of the analysis of algorithms consists in selecting them judiciously. We usually need induction to prove that a chosen loop invariant holds after each iteration of a loop, and usually we also need the pre-condition as an assumption in this proof.

An implicit pre-condition of all the algorithms in this section is that the numbers are in $\mathbb{Z} = \{\ldots, -2, -1, 0, 1, 2, \ldots\}$.

1.3.1 *Division algorithm*

We analyze the algorithm for integer division, algorithm 1.1. Note that the q and r returned by the division algorithm are usually denoted as $\text{div}(x, y)$ (the *quotient*) and $\text{rem}(x, y)$ (the *remainder*), respectively.

Algorithm 1.1 Division

Pre-condition: $x \geq 0 \wedge y > 0$

1: $q \leftarrow 0$
2: $r \leftarrow x$
3: **while** $y \leq r$ **do**
4: $\quad r \leftarrow r - y$
5: $\quad q \leftarrow q + 1$
6: **end while**
7: **return** q, r

Post-condition: $x = (q \cdot y) + r \wedge 0 \leq r < y$

We propose the following assertion as the loop invariant:

$$x = (q \cdot y) + r \wedge r \geq 0. \tag{1.4}$$

We show that (1.4) holds after each iteration of the loop. Basis case (i.e., zero iterations of the loop—we are just before line 3 of the algorithm): $q = 0, r = x$, so $x = (q \cdot y) + r$ and since $x \geq 0$ and $r = x$, $r \geq 0$.

Induction step: suppose $x = (q \cdot y) + r \wedge r \geq 0$ and we go once more through the loop, and let q', r' be the new values of q, r, respectively (computed in lines 4 and 5 of the algorithm). Since we executed the loop one more time it follows that $y \leq r$ (this is the condition checked for in line 3 of the algorithm), and since $r' = r - y$, we have that $r' \geq 0$. Thus,

$$x = (q \cdot y) + r = ((q + 1) \cdot y) + (r - y) = (q' \cdot y) + r',$$

and so q', r' still satisfy the loop invariant (1.4).

Now we use the loop invariant to show that (if the algorithm terminates) the post-condition of the division algorithm holds, *if* the pre-condition holds. This is very easy in this case since the loop ends when it is no longer true that $y \leq r$, i.e., when it is true that $r < y$. On the other hand, (1.4) holds after each iteration, and in particular the last iteration. Putting together (1.4) and $r < y$ we get our post-condition, and hence partial correctness.

To show termination we use the least number principle (LNP). We need to relate some non-negative monotone decreasing sequence to the algorithm; just consider $r_0, r_1, r_2, \ldots$, where $r_0 = x$, and r_i is the value of r after the i-th iteration. Note that $r_{i+1} = r_i - y$. First, $r_i \geq 0$, because the algorithm enters the while loop only if $y \leq r$, and second, $r_{i+1} < r_i$, since $y > 0$. By LNP such a sequence "cannot go on for ever," (in the sense that the set $\{r_i | i = 0, 1, 2, \ldots\}$ is a subset of the natural numbers, and so it has a least element), and so the algorithm must terminate.

Thus we have shown full correctness of the division algorithm.

Problem 1.17. *Write a Python program that takes as input x and y, and outputs the intermediate values of q and r, and finally the quotient and remainder of the division of x by y.*

1.3.2 *Euclid's algorithm*

Given two positive integers a and b, their *greatest common divisor*, denoted as $\gcd(a, b)$, is the largest positive integer that divides them both. Euclid's algorithm, presented as algorithm 1.2, is a procedure for finding the greatest common divisor of two numbers. It is one of the oldest know algorithms—it appeared in Euclid's *Elements* (Book 7, Propositions 1 and 2) around 300 BC.

Algorithm 1.2 Euclid

Pre-condition: $a > 0 \wedge b > 0$

1: $m \leftarrow a$; $n \leftarrow b$; $r \leftarrow \text{rem}(m, n)$
2: **while** $(r > 0)$ **do**
3: $\quad m \leftarrow n$; $n \leftarrow r$; $r \leftarrow \text{rem}(m, n)$
4: **end while**
5: **return** n

Post-condition: $n = \gcd(a, b)$

Note that to compute $\text{rem}(n, m)$ in lines 1 and 3 of Euclid's algorithm we need to use algorithm 1.1 (the division algorithm) as a subroutine; this is a typical "composition" of algorithms. Also note that lines 1 and 3 are executed from left to right, so in particular in line 3 we first do $m \leftarrow n$, then $n \leftarrow r$ and finally $r \leftarrow \text{rem}(m, n)$. This is important for the algorithm to work correctly.

To prove the correctness of Euclid's algorithm we are going to show that after each iteration of the while loop the following assertion holds:

$$m > 0, n > 0 \text{ and } \gcd(m, n) = \gcd(a, b), \tag{1.5}$$

that is, (1.5) is our loop invariant. We prove this by induction on the number of iterations. Basis case: after zero iterations (i.e., just before the while loop starts—so after executing line 1 and before executing line 2) we have that $m = a > 0$ and $n = b > 0$, so (1.5) holds trivially. Note that $a > 0$ and $b > 0$ by the pre-condition.

For the induction step, suppose $m, n > 0$ and $\gcd(a, b) = \gcd(m, n)$, and we go through the loop one more time, yielding m', n'. We want to show that $\gcd(m, n) = \gcd(m', n')$. Note that from line 3 of the algorithm we see that $m' = n, n' = r = \text{rem}(m, n)$, so in particular $m' = n > 0$ and $n' = r = \text{rem}(m, n) > 0$ since if $r = \text{rem}(m, n)$ were zero, the loop would have terminated (and we are assuming that we are going through the loop one more time). So it is enough to prove the assertion in Problem 1.18.

Problem 1.18. *Show that for all* $m, n > 0$, $\gcd(m, n) = \gcd(n, \mathrm{rem}(m, n))$.

Now the correctness of Euclid's algorithm follows from (1.5), since the algorithm stops when $r = \mathrm{rem}(m, n) = 0$, so $m = q \cdot n$, and so $\gcd(m, n) = n$.

Problem 1.19. *Show that Euclid's algorithm terminates.*

Problem 1.20. *Do you have any ideas how to speed-up Euclid's algorithm?*

Problem 1.21. *Modify Euclid's algorithm so that given integers* m, n *as input, it outputs integers* a, b *such that* $am + bn = g = \gcd(m, n)$. *This is called the* extended Euclid's algorithm.

(a) *Use the LNP to show that if* $g = \gcd(m, n)$, *then there exist* a, b *such that* $am + bn = g$.
(b) *Design Euclid's extended algorithm, and prove its correctness.*
(c) *The usual Euclid's extended algorithm has a running time polynomial in* $\min\{m, n\}$; *show that this is the running time of your algorithm, or modify your algorithm so that it runs in this time.*

Problem 1.22. *Write a Python program that implements Euclid's extended algorithm. Then perform the following experiment: run it on a random selection of inputs of a given size, for sizes bounded by some parameter* N; *compute the average number of steps of the algorithm for each input size* $n \leq N$, *and use* `gnuplot`[2] *to plot the result. What does* $f(n)$*—which is the "average number of steps" of Euclid's extended algorithm on input size* n*—look like? Note that size is not the same as value; inputs of size* n *are inputs with a binary representation of* n *bits.*

1.3.3 *Palindromes algorithm*

Algorithm 1.3 tests strings for *palindromes*, which are strings that read the same backwards as forwards, for example, `madamimadam` or `racecar`.

Let the loop invariant be: after the k-th iteration, $i = k + 1$ and for all j such that $1 \leq j \leq k$, $A[j] = A[n - j + 1]$. We prove that the loop invariant holds by induction on k. Basis case: before any iterations take place, i.e., after zero iterations, there are no j's such that $1 \leq j \leq 0$, so the

[2]Gnuplot is a command-line driven graphing utility with versions for most platforms. If you do not already have it, you can download it from `http://www.gnuplot.info` . If you prefer, you can use any other plotting utility.

Algorithm 1.3 Palindromes

Pre-condition: $n \geq 1 \wedge A[1 \ldots n]$ is a character array

```
1: i ← 1
2: while (i ≤ ⌊n/2⌋) do
3:     if (A[i] ≠ A[n − i + 1]) then
4:         return F
5:         i ← i + 1
6:     end if
7: end while
8: return T
```

Post-condition: return T iff A is a palindrome

---

second part of the loop invariant is (vacuously) true. The first part of the loop invariant holds since i is initially set to 1.

Induction step: we know that after k iterations, $A[j] = A[n-j+1]$ for all $1 \leq j \leq k$; after one more iteration we know that $A[k+1] = A[n-(k+1)+1]$, so the statement follows for all $1 \leq j \leq k+1$. This proves the loop invariant.

Problem 1.23. *Using the loop invariant argue the partial correctness of the palindromes algorithm. Show that the algorithm for palindromes always terminates.*

In is easy to manipulate strings in Python; a segment of a string is called a *slice*. Consider the word `palindrome`; if we set the variables `s` to this word,

```
s = 'palindrome'
```

then we can access different slices as follows:

```
print s[0:5]      palin
print s[5:10]     drome
print s[5:]       drome
print s[2:8:2]    lnr
```

where the notation `[i:j]` means the segment of the string starting from the i-th character (and we always start counting at zero!), to the j-th character, including the first but excluding the last. The notation `[i:]` means from the i-th character, all the way to the end, and `[i:j:k]` means starting from the i-th character to the j-th (again, not including the j-th itself), taking every k-th character.

One way to understand the string delimiters is to write the indices "in between" the numbers, as well as at the beginning and at the end. For example

$$_0\mathtt{p}_1\mathtt{a}_2\mathtt{l}_3\mathtt{i}_4\mathtt{n}_5\mathtt{d}_6\mathtt{r}_7\mathtt{o}_8\mathtt{m}_9\mathtt{e}_{10}$$

and to notice that a slice `[i:j]` contains all the symbols between index i and index j.

Problem 1.24. *Using Python's inbuilt facilities for manipulating slices of strings, write a succinct program that checks whether a given string is a palindrome.*

1.3.4 *Further examples*

Problem 1.25. *Give an algorithm which on the input "a positive integer n," outputs "yes" if $n = 2^k$ (i.e., n is a power of 2), and "no" otherwise. Prove that your algorithm is correct.*

Problem 1.26. *What does algorithm 1.4 compute? Prove your claim.*

Algorithm 1.4 Problem 1.26

1: $x \leftarrow m$; $y \leftarrow n$; $z \leftarrow 0$
2: **while** $(x \neq 0)$ **do**
3:     **if** $(\text{rem}(x, 2) = 1)$ **then**
4:         $z \leftarrow z + y$
5:     **end if**
6:     $x \leftarrow \text{div}(x, 2)$
7:     $y \leftarrow y \cdot 2$
8: **end while**
9: **return** z

Problem 1.27. *What does algorithm 1.5 compute? Assume that a, b are positive integers (i.e., assume that the pre-condition is that $a, b > 0$). For which starting a, b does this algorithm terminate? In how many steps does it terminate, if it does terminate?*

The following two problems require some linear algebra[3]. We say that a set of vectors $\{v_1, v_2, \dots, v_n\}$ is *linearly independent* if $\sum_{i=1}^{n} c_i v_i = 0$ implies that $c_i = 0$ for all i, and that they *span* a vector space $V \subseteq \mathbb{R}^n$

[3] A great and accessible introduction to linear algebra can be found in [Halmos (1995)].

Algorithm 1.5 Problem 1.27

```
1: while (a > 0) do
2:     if (a < b) then
3:         (a, b) ← (2a, b − a)
4:     else
5:         (a, b) ← (a − b, 2b)
6:     end if
7: end while
```

---

if whenever $v \in V$, then there exist $c_i \in \mathbb{R}$ such that $v = \sum_{i=1}^{n} c_i v_i$. We denote this as $V = \text{span}\{v_1, v_2, \ldots, v_n\}$. A set of vectors $\{v_1, v_2, \ldots, v_n\}$ in $\mathbb{R}^n$ is a *basis* for a vector space $V \subseteq \mathbb{R}^n$ if they are linearly independent and span V. Let $x \cdot y$ denote the *dot-product* of two vectors, defined as $x \cdot y = (x_1, x_2, \ldots, x_n) \cdot (y_1, y_2, \ldots, y_n) = \sum_{i=1}^{n} x_i y_i$, and the *norm* of a vector x is defined as $\|x\| = \sqrt{x \cdot x}$. Two vectors x, y are *orthogonal* if $x \cdot y = 0$.

Problem 1.28. *Let $V \subseteq \mathbb{R}^n$ be a vector space, and $\{v_1, v_2, \ldots, v_n\}$ its basis. Consider algorithm 1.6 and show that it produces an orthogonal basis*

Algorithm 1.6 Gram-Schmidt

Pre-condition: $\{v_1, \ldots, v_n\}$ a basis for $\mathbb{R}^n$

1: $v_1^* \longleftarrow v_1$
2: **for** $i = 2, 3, \ldots, n$ **do**
3:     **for** $j = 1, 2, \ldots, (i-1)$ **do**
4:         $\mu_{ij} \longleftarrow (v_i \cdot v_j^*)/\|v_j^*\|^2$
5:     **end for**
6:     $v_i^* \longleftarrow v_i - \sum_{j=1}^{i-1} \mu_{ij} v_j^*$
7: **end for**

Post-condition: $\{v_1^*, \ldots, v_n^*\}$ an orthogonal basis for $\mathbb{R}^n$

$\{v_1^*, v_2^*, \ldots, v_n^*\}$ for the vector space V. In other words, show that $v_i^* \cdot v_j^* = 0$ when $i \neq j$, and that $\text{span}\{v_1, v_2, \ldots, v_n\} = \text{span}\{v_1^*, v_2^*, \ldots, v_n^*\}$. Justify why in line 4 of the algorithm we never divide by zero.

Problem 1.29. *Implement the Gram-Schmidt algorithm (algorithm 1.6) in Python, but with the following twist: instead of computing over $\mathbb{R}$, the real numbers, compute over $\mathbb{Z}_2$, the field of two elements, where addition and multiplication are defined as follows:*

+	0	1
0	0	1
1	1	0

·	0	1
0	0	0
1	0	1

In fact, this "twist" makes the implementation much easier, as you do not have to deal with the precision issues involved in implementing division operations over the field of real numbers.

Suppose that $\{v_1, v_2, \ldots, v_n\}$ are linearly independent vectors in $\mathbb{R}^n$. The *lattice* L spanned by these vectors is the set $\{\sum_{i=1}^{n} c_i v_i : c_i \in \mathbb{Z}\}$, i.e., L consists of linear combinations of the vectors $\{v_1, v_2, \ldots, v_n\}$ where the coefficients are limited to be integers.

Problem 1.30. *Suppose that $\{v_1, v_2\}$ span a lattice in $\mathbb{R}^2$. Consider algorithm 1.7 and show that it terminates and outputs a new basis $\{v_1, v_2\}$ for L where v_1 is the shortest vector in the lattice L, i.e., $\|v_1\|$ is as small as possible among all the vectors of L.*

Algorithm 1.7 Gauss lattice reduction in dimension 2

Pre-condition: $\{v_1, v_2\}$ are linearly independent in $\mathbb{R}^2$

```
 1: loop
 2:     if ||v2|| < ||v2|| then
 3:         swap v1 and v2
 4:     end if
 5:     m ←— ⌊v1 · v2/||v1||²⌉
 6:     if m = 0 then
 7:         return v1, v2
 8:     else
 9:         v2 ←— v2 − m v1
10:     end if
11: end loop
```

Based on the examples presented thus far it may appear that it is fairly clear to the naked eye whether an algorithm terminates or not, and the difficulty consists in coming up with a proof. But that is not the case.

Clearly, if we have a trivial algorithm consisting of a single while-loop, with the condition $i \geq 0$, and the body of the loop consists of the single command $i \longleftarrow i+1$, then we can immediately conclude that this while-loop will never terminate. But what about algorithm 1.8? Does it terminate?

Algorithm 1.8 Ulam's algorithm

Pre-condition: $a > 0$

$x \longleftarrow a$
while last three values of x not $4, 2, 1$ **do**
  if x is even **then**
    $x \longleftarrow x/2$
  else
    $x \longleftarrow 3x + 1$
  end if
end while

For example, if $a = 22$, then one can check that x takes on the following values: $22, 11, 34, 17, 52, 26, 13, 40, 20, 10, 5, 16, 8, \mathbf{4}, \mathbf{2}, \mathbf{1}$, and algorithm 1.8 terminates.

It is conjectured that regardless of the initial value of a, as long as a is a positive integer, algorithm 1.8 terminates. This conjecture is known as "Ulam's problem,"[4] No one has been able to prove that algorithm 1.8 terminates, and in fact proving termination would involve solving a difficult open mathematical problem.

Problem 1.31. *Write a Python program that takes a as input and computes and displays all the values of Ulam's problem until it sees* $4, 2, 1$ *at which point it stops. You have just written a program for which there is no proof of termination.*

1.3.5 *Recursion and fixed points*

So far we have proved the correctness of while-loops and for-loops, but there is another way of "looping" using *recursive* procedures, i.e., algorithms that "call themselves." We are going to see examples of such algorithms in the chapter on the divide and conquer method.

There is a robust theory of correctness of recursive algorithms based on fixed point theory, and in particular on Kleene's theorem (see appendix B, theorem B.39). We briefly illustrate this approach with an example. We are going to be using partial orders; all the necessary background can be found in appendix B, in section B.3. Consider the recursive algorithm 1.9.

[4]It is also called "Collatz Conjecture," "Syracuse Problem," "Kakutani's Problem," or "Hasse's Algorithm." While it is true that a rose by any other name would smell just as sweet, the preponderance of names shows that the conjecture is a very alluring

Algorithm 1.9 $F(x, y)$

```
1: if x = y then
2:     return y + 1
3: else
4:     F(x, F(x − 1, y + 1))
5: end if
```

To see how this algorithm works consider computing $F(4, 2)$. First in line 1 it is established that $4 \neq 2$ and so we must compute $F(4, F(3, 3))$. We first compute $F(3, 3)$, recursively, so in line 1 it is now established that $3 = 3$, and so in line 2 y is set to 4 and that is the value returned, i.e., $F(3, 3) = 4$, so now we can go back and compute $F(4, F(3, 3)) = F(4, 4)$, so again, recursively, we establish in line 1 that $4 = 4$, and so in line 2 y is set to 5 and this is the value returned, i.e., $F(4, 2) = 5$. On the other hand it is easy to see that

$$F(3, 5) = F(3, F(2, 6)) = F(3, F(2, F(1, 7))) = \cdots,$$

and this procedure never ends as x will never equal y. Thus F is not a *total* function, i.e., not defined on all $(x, y) \in \mathbb{Z} \times \mathbb{Z}$.

Problem 1.32. *What is the* domain of definition *of F as computed by algorithm 1.9? That is, the domain of F is $\mathbb{Z} \times \mathbb{Z}$, while the domain of definition is the largest subset $S \subseteq \mathbb{Z} \times \mathbb{Z}$ such that F is defined for all $(x, y) \in S$. We have seen already that $(4, 2) \in S$ while $(3, 5) \notin S$.*

We now consider three different functions, all given by algorithms that are not recursive: algorithms 1.10, 1.11 and 1.12, computing functions f_1, f_2 and f_3, respectively.

Algorithm 1.10 $f_1(x, y)$

```
if x = y then
    return y + 1
else
    return x + 1
end if
```

Functions f_1 has an interesting property: if we were to replace F in algorithm 1.9 with f_1 we would get back F. In other words, given algorithm 1.9,

mathematical problem.

if we were to replace line 4 with $f_1(x, f_1(x-1, y+1))$, and compute f_1 with the (non-recursive) algorithm 1.10 for f_1, then algorithm 1.9 thus modified would now be computing $F(x, y)$. Therefore, we say that the functions f_1 is a *fixed point* of the recursive algorithm 1.9.

For example, recall the we have already shown that $F(4, 2) = 5$, using the recursive algorithm 1.9 for computing F. Replace line 4 in algorithm 1.9 with $f_1(x, f_1(x-1, y+1))$ and compute $F(4, 2)$ anew; since $4 \neq 2$ we go directly to line 4 where we compute $f_1(4, f_1(3, 3)) = f_1(4, 4) = 5$. Notice that this last computation was not recursive, as we computed f_1 directly with algorithm 1.10, and that we have obtained the same value.

Consider now f_2, f_3, computed by algorithms 1.11, 1.12, respectively.

Algorithm 1.11 $f_2(x, y)$

```
if x ≥ y then
        return x + 1
else
        return y − 1
end if
```

Algorithm 1.12 $f_3(x, y)$

```
if x ≥ y ∧ (x − y is even) then
        return x + 1
end if
```

Notice that in algorithm 1.12, if it is not the case that $x \geq y$ and $x - y$ is even, then the output is undefined. Thus f_3 is a partial function, and if $x < y$ or $x - y$ is odd, then (x, y) is not in its domain of definition.

Problem 1.33. *Prove that f_1, f_2, f_3 are all fixed points of algorithm 1.9.*

The function f_3 has one additional property. For every pair of integers x, y such that $f_3(x, y)$ is defined, that is $x \geq y$ and $x-y$ is even, both $f_1(x, y)$ and $f_2(x, y)$ are also defined and have the same value as $f_3(x, y)$. We say that f_3 is *less defined than or equal to* f_1 and f_2, and write $f_3 \sqsubseteq f_1$ and $f_3 \sqsubseteq f_2$; that is, we have defined (informally) a partial order on functions $f : \mathbb{Z} \times \mathbb{Z} \longrightarrow \mathbb{Z} \times \mathbb{Z}$.

Problem 1.34. *Show that $f_3 \sqsubseteq f_1$ and $f_3 \sqsubseteq f_2$. Recall the notion of a domain of definition introduced in problem 1.32. Let S_1, S_2, S_3 be the* *domains of definition of* f_1, f_2, f_3, *respectively. You must show that* $S_3 \subseteq S_1$ *and* $S_3 \subseteq S_2$.

It can be shown that f_3 has this property, not only with respect to f_1 and f_2, but also with respect to all fixed points of algorithm 1.9. Moreover, $f_3(x, y)$ is the only function having this property, and therefore f_3 is said to be the *least (defined) fixed point of* algorithm 1.9. It is an important application of Kleene's theorem (theorem B.39) that every recursive algorithm has a unique fixed point.

1.3.6 *Formal verification*

The proofs of correctness we have been giving thus far are considered to be "informal" mathematical proofs. There is nothing wrong with an informal proof, and in many cases such a proof is all that is necessary to convince oneself of the validity of a small "code snippet." However, there are many circumstances where extensive formal code validation is necessary; in that case, instead of an informal paper-and-pencil type of argument, we often employ computer assisted software verification. For example, the US Food and Drug Administration requires software certification in cases where medical devices are dependent on software for their effective and safe operation. When formal verification is required everything has to be stated explicitly, in a formal language, and proven painstakingly line by line. In this section we give an example of such a procedure.

Let $\{\alpha\}P\{\beta\}$ mean that if formula α is true before execution of P, P is executed and terminates, then formula β will be true, i.e., α, β, are the precondition and postcondition of the program P, respectively. They are usually given as formulas in some formal theory, such as first order logic over some language $\mathcal{L}$. We assume that the language is Peano Arithmetic; see Appendix C.

Using a finite set of rules for program verification, we want to show that $\{\alpha\}P\{\beta\}$ holds, and conclude that the program is correct *with respect to the specification* α, β. As our example is small, we are going to use a limited set of rules for program verification, given in figure 1.3

The "If" rule is saying the following: suppose that it is the case that $\{\alpha \wedge \beta\}P_1\{\gamma\}$ and $\{\alpha \wedge \neg\beta\}P_2\{\gamma\}$. This means that P_1 is (partially) correct with respect to precondition $\alpha \wedge \beta$ and postcondition γ, while P_2 is (partially) correct with respect to precondition $\alpha \wedge \neg\beta$ and postcondition γ. Then the program "**if** β **then** P_1 **else** P_2" is (partially) correct with

Consequence left and right

$$\frac{\{\alpha\}P\{\beta\} \qquad (\beta \rightarrow \gamma)}{\{\alpha\}P\{\gamma\}} \qquad \frac{(\gamma \rightarrow \alpha) \qquad \{\alpha\}P\{\beta\}}{\{\gamma\}P\{\beta\}}$$

Composition and assignment

$$\frac{\{\alpha\}P_1\{\beta\} \qquad \{\beta\}P_2\{\gamma\}}{\{\alpha\}P_1P_2\{\gamma\}} \qquad \frac{x := t}{\{\alpha(t)\}x := t\{\alpha(x)\}}$$

If

$$\frac{\{\alpha \wedge \beta\}P_1\{\gamma\} \qquad \{\alpha \wedge \neg\beta\}P_2\{\gamma\}}{\{\alpha\} \textbf{ if } \beta \textbf{ then } P_1 \textbf{ else } P_2 \ \{\gamma\}}$$

While

$$\frac{\{\alpha \wedge \beta\}P\{\alpha\}}{\{\alpha\} \textbf{ while } \beta \textbf{ do } P \ \{\alpha \wedge \neg\beta\}}$$

Fig. 1.3 A small set of rules for program verification.

respect to precondition α and postcondition γ because if α holds before it executes, then either β or $\neg\beta$ must be true, and so either P_1 or P_2 executes, respectively, giving us γ in both cases.

The "While" rule is saying the following: suppose it is the case that $\{\alpha \wedge \beta\}P\{\alpha\}$. This means that P is (partially) correct with respect to precondition $\alpha \wedge \beta$ and postcondition α. Then the program "**while** β **do** P" is (partially) correct with respect to precondition α and postcondition $\alpha \wedge \neg\beta$ because if α holds before it executes, then either β holds in which case the while-loop executes once again, with $\alpha \wedge \beta$ holding, and so α still holds after P executes, or β is false, in which case $\neg\beta$ is true and the loop terminates with $\alpha \wedge \neg\beta$.

As an example, we verify which computes $y = A \cdot B$. Note that in algorithm 1.13, which describes the program that computes $y = A \cdot B$, we use "=" instead of the usual "←" since we are now proving the correctness of an actual program, rather than its representation in pseudo-code.

We want to show:

$$\{B \geq 0\}\texttt{mult(A,B)}\{y = AB\} \tag{1.6}$$

Each pass through the while loop adds a to y, but $a \cdot b$ decreases by a because b is decremented by 1. Let the loop invariant be: $(y + (a \cdot b) = A \cdot B) \wedge b \geq 0$. To save space, write tu instead of $t \cdot u$. Let $t \geq u$ abbreviate the $\mathcal{L}_A$-formula $\exists x(t = u + x)$, and let $t \leq u$ abbreviate $u \geq t$.

Algorithm 1.13 `mult(A,B)`

Pre-condition: $B \geq 0$

$a = A;$
$b = B;$
$y = 0;$
while $b > 0$ **do**
  $y = y + a;$
  $b = b - 1;$
end while

Post-condition: $y = A \cdot B$

1 $\{y + a(b-1) = AB \wedge (b-1) \geq 0\}$`b=b-1;`$\{y + ab = AB \wedge b \geq 0\}$
assignment
2 $\{(y+a)+a(b-1) = AB \wedge (b-1) \geq 0\}$`y=y+a;`$\{y+a(b-1) = AB \wedge (b-1) \geq 0\}$
assignment
3 $(y + ab = AB \wedge b - 1 \geq 0) \rightarrow ((y + a) + a(b-1) = AB \wedge b - 1 \geq 0)$
theorem
4 $\{y + ab = AB \wedge b - 1 \geq 0\}$`y=y+a;`$\{y + a(b-1) = AB \wedge b - 1 \geq 0\}$
consequence left 2 and 3
5 $\{y + ab = AB \wedge b - 1 \geq 0\}$`y=y+a;b=b-1;`$\{y + ab = AB \wedge b \geq 0\}$
composition on 4 and 1
6 $(y + ab = AB) \wedge b \geq 0 \wedge b > 0 \rightarrow (y + ab = AB) \wedge b - 1 \geq 0$
theorem
7 $\{(y + ab = AB) \wedge b \geq 0 \wedge b > 0\}$`y=y+a; b=b-1;`$\{y + ab = AB \wedge b \geq 0\}$
consequence left 5 and 6

8 $\{(y+ab = AB) \wedge b \geq 0\}$
```
while (b>0)
    y=y+a;
    b=b-1;
```
$\{y+ab = AB \wedge b \geq 0 \wedge \neg(b > 0)\}$

while on 7
9 $\{(0 + ab = AB) \wedge b \geq 0\}$ `y=0;` $\{(y + ab = AB) \wedge b \geq 0\}$
assignment

10 $\{(0+ab = AB) \wedge b \geq 0\}$
```
y=0;
while (b>0)
    y=y+a;
    b=b-1;
```
$\{y+ab = AB \wedge b \geq 0 \wedge \neg(b > 0)\}$

composition on 9 and 8
11 $\{(0 + aB = AB) \wedge B \geq 0\}$ `b=B;` $\{(0 + ab = AB) \wedge b \geq 0\}$

assignment

```
                              b=B;
                              y=0;
```
12 $\{(0+aB=AB)\wedge B\geq 0\}$ `while (b>0)` $\{y+ab=AB\wedge b\geq 0\wedge\neg(b>0)\}$
```
                                  y=y+a;
                                  b=b-1;
```

composition on 11 and 10

13 $\{(0+AB=AB)\wedge B\geq 0\}$ `a=A;` $\{(0+aB=AB)\wedge B\geq 0\}$

assignment

14 $\{(0+AB=AB)\wedge B\geq 0\}$ `mult(A,B)` $\{y+ab=AB\wedge b\geq 0\wedge\neg(b>0)\}$

composition on 13 and 12

15 $B\geq 0\rightarrow((0+AB=AB)\wedge B\geq 0)$

theorem

16 $(y+ab=AB\wedge b\geq 0\wedge\neg(b>0))\rightarrow y=AB$

theorem

17 $\{B\geq 0\}$ `mult(A,B)` $\{y+ab=AB\wedge b\geq 0\wedge\neg(b>0)\}$

consequence left on 15 and 14

18 $\{B\geq 0\}$ `mult(A,B)` $\{y=AB\}$

consequence right on 16 and 17

Problem 1.35. *The following is a project, rather than an exercise. Give formal proofs of correctness of the division algorithm and Euclid's algorithm (algorithms 1.1 and 1.2). To give a complete proof you will need to use Peano Arithmetic, which is a formalization of number theory—exactly what is needed for these two algorithms. The details of Peano Arithmetic are given in Appendix C.*

1.4 Stable marriage

The method of "pairwise comparisons" was first described by Marquis de Condorcet in 1785. Today rankings based on pairwise comparisons are pervasive: scheduling of processes, online shopping and dating websites, to name just a few. We end this chapter with an elegant application known as the "stable marriage problem," which has been used since the 1960s for the college admission process and for matching interns with hospitals.

An instance of the *stable marriage problem* of size n consists of two disjoint finite sets of equal size; a set of *boys* $B=\{b_1,b_2,\ldots,b_n\}$, and a set

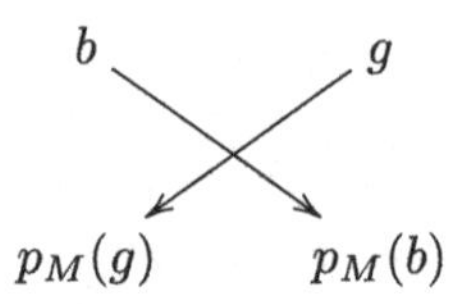

Fig. 1.4 A blocking pair: b and g prefer each other to their partners $p_M(b)$ and $p_M(g)$.

of *girls* $G = \{g_1, g_2, \ldots, g_n\}$. Let "$<_i$" denote the ranking of boy b_i; that is, $g <_i g'$ means that boy b_i prefers g over g'. Similarly, "$<^j$" denotes the ranking of girl g_j. Each boy b_i has such a ranking (linear ordering) $<_i$ of G which reflects his preference for the girls that he wants to marry. Similarly each girl g_j has a ranking (linear ordering) $<^j$ of B which reflects her preference for the boys she would like to marry.

A *matching* (or *marriage*) M is a 1-1 correspondence between B and G. We say that b and g are *partners* in M if they are matched in M and write $p_M(b) = g$ and also $p_M(g) = b$. A matching M is *unstable* if there is a pair (b, g) from $B \times G$ such that b and g are not partners in M but b prefers g to $p_M(b)$ and g prefers b to $p_M(g)$. Such a pair (b, g) is said to *block* the matching M and is called a *blocking pair* for M (see figure 1.4). A matching M is stable if it contains no blocking pair.

We are going to present an algorithm due to Gale and Shapley ([Gale and Shapley (1962)]) that outputs a stable marriage for any input B, G, regardless of the ranking.

The matching M is produced in stages M_s so that b_t always has a partner at the end of stage s, where $s \geq t$. However, the partners of b_t do not get better, i.e., $p_{M_t}(b_t) \leq_t p_{M_{t+1}}(b_t) \leq_t \cdots$. On the other hand, for each $g \in G$, if g has a partner at stage t, then g will have a partner at each stage $s \geq t$ and the partners do not get worse, i.e., $p_{M_t}(g) \geq^t p_{M_{t+1}}(g) \geq^t \ldots$. Thus, as s increases, the partners of b_t become less preferable and the partners of g become more preferable.

At the end of stage s, assume that we have produced a matching

$$M_s = \{(b_1, g_{1,s}), \ldots, (b_s, g_{s,s})\},$$

where the notation $g_{i,s}$ means that $g_{i,s}$ is the partner of boy b_i after the end of stage s.

We will say that partners in M_s are *engaged*. The idea is that at stage $s+1$, b_{s+1} will try to get a partner by *proposing* to the girls in G in his order of preference. When b_{s+1} proposes to a girl g_j, g_j accepts his proposal if either g_j is not currently engaged or is currently engaged to a less preferable boy b, i.e., $b_{s+1} <^j b$. In the case where g_j prefers b_{s+1} over her current partner b, then g_j breaks off the engagement with b and b then has to search for a new partner.

Algorithm 1.14 Gale-Shapley

Stage 1: At stage 1, b_1 chooses the first girl g in his preference list and we set $M_1 = \{(b_1, g)\}$.
Stage s + 1:
$M \longleftarrow M_s$
$b^* \longleftarrow b_{s+1}$
Then b^* proposes to the girls in order of his preference until one accepts; girl g will accept the proposal as long as she is either not engaged or prefers b^* to her current partner $p_M(g)$.
Then we add (b^*, g) to M and proceed according to one of the following two cases:
(i) If g was not engaged, then we terminate the procedure and set
$M_{s+1} \longleftarrow M \cup \{(b^*, g)\}$.
(ii) If g was engaged to b, then we set
$M \longleftarrow (M - \{(b, g)\}) \cup \{(b^*, g)\}$
$b^* \longleftarrow b$
and repeat.

Problem 1.36. *Show that each b need propose at most once to each g.*

From problem 1.36 we see that we can make each boy keep a bookmark on his list of preference, and this bookmark is only moving forward. When a boy's turn to choose comes, he starts proposing from the point where his bookmark is, and by the time he is done, his bookmark moved only forward. Note that at stage $s + 1$ each boy's bookmark cannot have moved beyond the girl number s on the list without choosing someone (after stage s only s girls are engaged). As the boys take turns, each boy's bookmark is advancing, so some boy's bookmark (among the boys in $\{b_1, \ldots, b_{s+1}\}$) will advance eventually to a point where he must choose a girl.

The discussion in the above paragraph shows that stage $s + 1$ in algorithm 1.14 must end. The concern here was that case (ii) of stage $s + 1$ might end up being circular. But the fact that the bookmarks are advancing shows that this is not possible.

Furthermore, this gives an upper bound of $(s+1)^2$ steps at stage $(s+1)$ in the procedure. This means that there are n stages, and each stage takes $O(n^2)$ steps, and hence algorithm 1.14 takes $O(n^3)$ steps altogether. The question, of course, is what do we mean by a step? Computers operate on binary strings, yet here the implicit assumption is that we compare numbers and access the lists of preferences in a single step. But the cost of these operations is negligible when compared to our idealized running time, and so we allow ourselves this poetic license to bound the overall running time.

Problem 1.37. *Show that there is exactly one girl that was not engaged at stage s but is engaged at stage $(s+1)$ and that, for each girl g_j that is engaged in M_s, g_j will be engaged in M_{s+1} and that $p_{M_{s+1}}(g_j) <^j p_{M_s}(g_j)$. (Thus, once g_j becomes engaged, she will remain engaged and her partners will only gain in preference as the stages proceed.)*

Problem 1.38. *Suppose that $|B| = |G| = n$. Show that at the end of stage n, M_n will be a stable marriage.*

We say that a matching (b, g) is *feasible* if there exists a stable matching in which b, g are partners. We say that a matching is boy-*optimal* if every boy is paired with his highest ranked feasible partner. We say that a matching is boy-*pessimal* if every boy is paired with his lowest ranking feasible partner. Similarly, we define girl-*optimal/pessimal.*

Problem 1.39. *Show that our version of the algorithm produces a boy-optimal and girl-pessimal stable matching.*

Problem 1.40. *Implement the stable marriage problem algorithm in Python. Let the input be given as a text file containing, on each line, the preference list for each boy and girl.*

1.5 Answers to selected problems

Problem 1.2. Basis case: $n = 1$, then $1^3 = 1^2$. For the induction step:

$$\begin{aligned}
&(1+2+3+\cdots+n+(n+1))^2 \\
&= (1+2+3+\cdots+n)^2 + 2(1+2+3+\cdots+n)(n+1) + (n+1)^2
\end{aligned}$$

and by the induction hypothesis,

$$
\begin{aligned}
&= (1^3 + 2^3 + 3^3 + \cdots + n^3) + 2(1 + 2 + 3 + \cdots + n)(n + 1) + (n + 1)^2 \\
&= (1^3 + 2^3 + 3^3 + \cdots + n^3) + 2\frac{n(n + 1)}{2}(n + 1) + (n + 1)^2 \\
&= (1^3 + 2^3 + 3^3 + \cdots + n^3) + n(n + 1)^2 + (n + 1)^2 \\
&= (1^3 + 2^3 + 3^3 + \cdots + n^3) + (n + 1)^3
\end{aligned}
$$

Problem 1.3. It is important to interpret the statement of the problem correctly: when it says that one square is missing, it means that *any* square may be missing. So the basis case is: given a 2×2 square, there are four possible ways for a square to be missing; but in each case, the remaining squares form an "L." These four possibilities are drawn in figure 1.5.

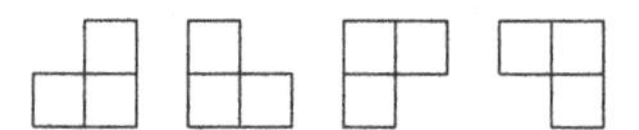

Fig. 1.5 The four different "L" shapes.

Suppose the claim holds for n, and consider a square of size $2^{n+1} \times 2^{n+1}$. Divide it into four quadrants of equal size. No matter which square we choose to be missing, it will be in one of the four quadrants; that quadrant can be filled with "L" shapes (i.e., shapes of the form given by figure 1.5) by induction hypothesis. As to the remaining three quadrants, put an "L" in them in such a way that it straddles all three of them (the "L" wraps around the center staying in those three quadrants). The remaining squares of each quadrant can now be filled with "L" shapes by induction hypothesis.

Problem 1.4. Since $\forall n(P(n) \to P(n+1)) \to (\forall n \geq k)(P(n) \to P(n+1))$, then $(1.2) \Rightarrow (1.2')$. On the other hand, $(1.2') \not\Rightarrow (1.2)$.

Problem 1.5. The basis case is $n = 1$, and it is immediate. For the induction step, assume the equality holds for exponent n, and show that it holds for exponent $n + 1$:

$$
\begin{pmatrix} 1 & 1 \\ 1 & 0 \end{pmatrix}^n \begin{pmatrix} 1 & 1 \\ 1 & 0 \end{pmatrix} = \begin{pmatrix} f_{n+1} & f_n \\ f_n & f_{n-1} \end{pmatrix} \begin{pmatrix} 1 & 1 \\ 1 & 0 \end{pmatrix} = \begin{pmatrix} f_{n+1} + f_n & f_{n+1} \\ f_n + f_{n-1} & f_n \end{pmatrix}
$$

The right-most matrix can be simplified using the definition of Fibonacci numbers to be as desired.

Problem 1.7. $m|n$ iff $n = km$, so show that $f_m | f_{km}$ by induction on k. If $k = 1$, there is nothing to prove. Otherwise, $f_{(k+1)m} = f_{km+m}$. Now, using a separate inductive argument, show that for $y \geq 1$, $f_{x+y} =$

$f_y f_{x+1} + f_{y-1} f_x$, and finish the proof. To show this last statement, let $y = 1$, and note that $f_y f_{x+1} + f_{y-1} f_x = f_1 f_{x+1} + f_0 f_x = f_{x+1}$. Assume now that $f_{x+y} = f_y f_{x+1} + f_{y-1} f_x$ holds. Consider

$$\begin{aligned} f_{x+(y+1)} &= f_{(x+y)+1} = f_{(x+y)} + f_{(x+y)-1} = f_{(x+y)} + f_{x+(y-1)} \\ &= (f_y f_{x+1} + f_{y-1} f_x) + (f_{y-1} f_{x+1} + f_{y-2} f_x) \\ &= f_{x+1}(f_y + f_{y-1}) + f_x(f_{y-1} + f_{y-2}) \\ &= f_{x+1} f_{y+1} + f_x f_y. \end{aligned}$$

Problem 1.8. Note that this is almost the *Fundamental Theorem of Arithmetic*; what is missing is the fact that up to reordering of primes this representation is unique. The proof of this can be found in appendix A, theorem A.2.

Problem 1.9. Let our assertion P(n) be: the minimal number of breaks to break up a chocolate bar of n squares is $(n-1)$. Note that this says that $(n-1)$ breaks are sufficient, and $(n-2)$ are not. Basis case: only one square requires no breaks. Induction step: Suppose that we have $m+1$ squares. No matter how we break the bar into two smaller pieces of a and b squares each, $a + b = m + 1$.

By induction hypothesis, the "a" piece requires $a - 1$ breaks, and the "b" piece requires $b - 1$ breaks, so together the number of breaks is

$$(a-1) + (b-1) + \boxed{\mathbf{1}} = a + b - 1 = m + 1 - 1 = m,$$

and we are done. Note that the **1** in the box comes from the initial break to divide the chocolate bar into the "a" and the "b" pieces.

So the "boring" way of breaking up the chocolate (first into rows, and then each row separately into pieces) is in fact optimal.

Problem 1.10. Let IP be: $[\mathrm{P}(0) \wedge (\forall n)(\mathrm{P}(n) \rightarrow \mathrm{P}(n+1))] \rightarrow (\forall m)\mathrm{P}(m)$ (where n, m range over natural numbers), and let LNP: *Every non-empty subset of the natural numbers has a least element.* These two principles are equivalent, in the sense that one can be shown from the other. Indeed:

LNP⇒IP: Suppose we have $[\mathrm{P}(0) \wedge (\forall n)(\mathrm{P}(n) \rightarrow \mathrm{P}(n+1))]$, but that it is *not* the case that $(\forall m)\mathrm{P}(m)$. Then, the set S of m's for which P(m) is false is non-empty. By the LNP we know that S has a least element. We know this element is not 0, as P(0) was assumed. So this element can be expressed as $n + 1$ for some natural number n. But since $n + 1$ is the least such number, P(n) must hold. This is a contradiction as we assumed that $(\forall n)(\mathrm{P}(n) \rightarrow \mathrm{P}(n+1))$, and here we have an n such that P(n) but not P$(n+1)$.

IP⇒LNP: Suppose that S is a non-empty subset of the natural numbers. Suppose that it does not have a least element; let P(n) be the following assertion "all elements up to and including n are not in S." We know that P(0) must be true, for otherwise 0 would be in S, and it would then be the least element (by definition of 0). Suppose P(n) is true (so none of $\{0, 1, 2, \ldots, n\}$ is in S). Suppose P($n + 1$) were false: then $n + 1$ would necessarily be in S (as we know that none of $\{0, 1, 2, \ldots, n\}$ is in S), and thereby $n + 1$ would be the smallest element in S. So we have shown $[\mathrm{P}(0) \wedge (\forall n)(\mathrm{P}(n) \rightarrow \mathrm{P}(n + 1))]$. By IP we can therefore conclude that $(\forall m)\mathrm{P}(m)$. But this means that S is empty. Contradiction. Thus S must have a least element.

IP⇒CIP: For this direction we use the LNP which we just showed equivalent to the IP. Suppose that we have IP; assume that $P(0)$ and $\forall n((\forall i \leq n)P(i) \rightarrow P(n+1))$. We want to show that $\forall nP(n)$, so we prove this with the IP: the basis case, $P(0)$, is given. To show $\forall j(P(j) \rightarrow P(j+1))$ suppose that it does not hold; then there exists a j such that $P(j)$ and $\neg P(j)$; let j be the smallest such j; one exists by the LNP, and $j \neq 0$ by what is given. So $P(0), P(1), P(2), \ldots, P(j)$ but $\neg P(j + 1)$. But this contradicts $\forall n((\forall i \leq n)P(i) \rightarrow P(n + 1))$, and so it is not possible. Hence $\forall j(P(j) \rightarrow P(j + 1))$ and so by the IP we have $\forall nP(n)$ and hence we have the CIP.

The last direction, CIP⇒IP, follows directly from the fact that CIP has a "stronger" induction step.

Problem 1.11. We use the example in figure 1.1. Suppose that we want to obtain the tree from the infix (2164735) and prefix (1234675) encodings: from the prefix encoding we know that 1 is the root, and thus from the infix encoding we know that the left sub-tree has the infix encoding 2, and so prefix encoding 2, and the right sub-tree has the infix encoding 64735 and so prefix encoding 34675, and we proceed recursively.

Problem 1.13. Consider the following invariant: the sum S of the numbers currently in the set is odd. Now we prove that this invariant holds. Basis case: $S = 1+2+\cdots+2n = n(2n+1)$ which is odd. Induction step: assume S is odd, let S' be the result of one more iteration, so

$$S' = S + |a - b| - a - b = S - 2\min(a, b),$$

and since $2\min(a, b)$ is even, and S was odd by the induction hypothesis, it follows that S' must be odd as well. At the end, when there is just one number left, say x, $S = x$, so x is odd.

Problem 1.14. To solve this problem we must provide both an algorithm and an invariant for it. The algorithm works as follows: initially divide the club into any two groups. Let H be the total sum of enemies that each member has in his own group. Now repeat the following loop: while there is an m which has at least two enemies in his own group, move m to the other group (where m must have at most one enemy). Thus, when m switches houses, H decreases. Here the invariant is "H decreases monotonically." Now we know that a sequence of positive integers cannot decrease for ever, so when H reaches its absolute minimum, we obtain the required distribution.

Problem 1.15. At first, arrange the guests in any way; let H be the number of neighboring hostile pairs. We find an algorithm that reduces H whenever $H > 0$. Suppose $H > 0$, and let (A, B) be a hostile couple, sitting side-by-side, in the clockwise order A, B. Traverse the table, clockwise, until we find another couple (A', B') such that A, A' and B, B' are friends. Such a couple must exist: there are $2n - 2 - 1 = 2n - 3$ candidates for A' (these are all the people sitting clockwise after B, which have a neighbor sitting next to them, again clockwise, and that neighbor is neither A nor B). As A has at least n friends (among people other than itself), out of these $2n - 3$ candidates, at least $n - 1$ of them are friends of A. If each of these friends had an enemy of B sitting next to it (again, going clockwise), then B would have at least n enemies, which is not possible, so there must be an A' friends with A so that the neighbor of A' (clockwise) is B' and B' is a friend of B; see figure 1.6.

Note that when $n = 1$ no one has enemies, and so this analysis is applicable when $n \geq 2$, in which case $2n - 3 \geq 1$.

Now the situation around the table is $\ldots, A, \boxed{B, \ldots, A'}, B', \ldots$. Reverse everyone in the box (i.e., mirror image the box), to reduce H by 1. Keep repeating this procedure while $H > 0$; eventually $H = 0$ (by the LNP), at which point there are no neighbors that dislike each other.

$$A, B, c_1, c_2, \ldots, c_{2n-3}, c_{2n-2}$$

Fig. 1.6 List of guests sitting around the table, in clockwise order, starting at A. We are interested in friends of A among $c_1, c_2, \ldots, c_{2n-3}$, to make sure that there is a neighbor to the right, and that neighbor is not A or B; of course, the table wraps around at c_{2n-2}, so the next neighbor, clockwise, of c_{2n-2} is A. As A has at most $n - 1$ enemies, A has at least n friends (not counting itself; self-love does not count as friendship). Those n friends of A are among the c's, but if we exclude c_{2n-2} it follows that A has at least $n-1$ friends among $c_1, c_2, \ldots, c_{2n-3}$. If the clockwise neighbor of c_i, $1 \leq i \leq 2n-3$, i.e., c_{i+1} was in each case an enemy of B, then, as B already has an enemy of A, it would follow that B has n enemies, which is not possible.

Problem 1.16. We partition the participants into the set E of even persons and the set O of odd persons. We observe that, during the hand shaking ceremony, the set O cannot change its parity. Indeed, if two odd persons shake hands, O decreases by 2. If two even persons shake hands, O increases by 2, and, if an even and an odd person shake hands, $|O|$ does not change. Since, initially, $|O| = 0$, the parity of the set is preserved.

Problem 1.18. First observe that if u divides x and y, then for any $a, b \in \mathbb{Z}$ u also divides $ax + by$. Thus, if $i|m$ and $i|n$, then

$$i|(m - qn) = r = \text{rem}(m, n).$$

So i divides both n and $\text{rem}(m,n)$, and so i has to be bounded by their greatest common divisor, i.e., $i \leq \gcd(n, \text{rem}(m,n))$. As this is true for every i, it is in particular true for $i = \gcd(m,n)$; thus $\gcd(m,n) \leq \gcd(n, \text{rem}(m,n))$. Conversely, suppose that $i|n$ and $i|\text{rem}(m,n)$. Then $i|m = qn + r$, so $i \leq \gcd(m,n)$, and again, $\gcd(n, \text{rem}(m,n))$ meets the condition of being such an i, so we have $\gcd(n, \text{rem}(m,n)) \leq \gcd(m,n)$. Both inequalities taken together give us $\gcd(m,n) = \gcd(n, \text{rem}(m,n))$.

Problem 1.19. Let r_i be r after the i-th iteration of the loop. Note that $r_0 = \text{rem}(m,n) = \text{rem}(a,b) \geq 0$, and in fact every $r_i \geq 0$ by definition of remainder. Furthermore:

$$r_{i+1} = \text{rem}(m', n') = \text{rem}(n, r) = \text{rem}(n, \text{rem}(m,n)) = \text{rem}(n, r_i) < r_i.$$

and so we have a decreasing, and yet non-negative, sequence of numbers; by the LNP this must terminate.

Problem 1.20. When $m < n$ then $\text{rem}(m,n) = m$, and so $m' = n$ and $n' = m$. Thus, when $m < n$ we execute one iteration of the loop only to swap m and n. In order to be more efficient, we could add line 2.5 in algorithm 1.2 saying **if** $(m < n)$ **then** swap(m,n).

Problem 1.21. (a) We show that if $d = \gcd(a,b)$, then there exist u, v such that $au + bv = d$. Let $S = \{ax + by | ax + by > 0\}$; clearly $S \neq \emptyset$. By LNP there exists a least $g \in S$. We show that $g = d$. Let $a = q \cdot g + r$, $0 \leq r < g$. Suppose that $r > 0$; then

$$r = a - q \cdot g = a - q(ax_0 + by_0) = a(1 - qx_0) + b(-qy_0).$$

Thus, $r \in S$, but $r < g$—contradiction. So $r = 0$, and so $g|a$, and a similar argument shows that $g|b$. It remains to show that g is greater than any other common divisor of a, b. Suppose $c|a$ and $c|b$, so $c|(ax_0 + by_0)$, and so $c|g$, which means that $c \leq g$. Thus $g = \gcd(a,b) = d$.

(b) Euclid's extended algorithm is algorithm 1.15. Note that in the algorithm, the assignments in line 1 and line 8 are evaluated left to right.

Algorithm 1.15 Extended Euclid's algorithm.

Pre-condition: $m > 0, n > 0$

1: $a \leftarrow 0;\ x \leftarrow 1;\ b \leftarrow 1;\ y \leftarrow 0;\ c \leftarrow m;\ d \leftarrow n$
2: **loop**
3: $\quad q \leftarrow \text{div}(c, d)$
4: $\quad r \leftarrow \text{rem}(c, d)$
5: $\quad$ **if** $r = 0$ **then**
6: $\qquad$ stop
7: $\quad$ **end if**
8: $\quad c \leftarrow d;\ d \leftarrow r;\ t \leftarrow x;\ x \leftarrow a;\ a \leftarrow t - qa;\ t \leftarrow y;\ y \leftarrow b;\ b \leftarrow t - qb$
9: **end loop**

Post-condition: $am + bn = d = \gcd(m, n)$

We can prove the correctness of algorithm 1.15 by using the following loop invariant which consists of four assertions:

$$am + bn = d, \quad xm + yn = c, \quad d > 0, \quad \gcd(c, d) = \gcd(m, n). \qquad \text{(LI)}$$

The basis case:

$$am + bn = 0 \cdot m + 1 \cdot n = n = d$$
$$xm + yn = 1 \cdot m + 0 \cdot n = m = c$$

both by line 1. Then $d = n > 0$ by pre-condition, and $\gcd(c, d) = \gcd(m, n)$ by line 1. For the induction step assume that the "primed" variables are the result of one more full iteration of the loop on the "un-primed" variables:

$$\begin{aligned} a'm + b'n &= (x - qa)m + (y - qb)n && \text{by line 8} \\ &= (xm - yn) - q(am + bn) && \\ &= c - qd && \text{by induction hypothesis} \\ &= r && \text{by lines 3 and 4} \\ &= d' && \text{by line 8} \end{aligned}$$

Then $x'm = y'n = am+bn = d = c'$ where the first equality is by line 8, the second by the induction hypothesis, and the third by line 8. Also, $d' = r$ by line 8, and the algorithm would stop in line 5 if $r = 0$; on the other hand, from line 4, $r = \text{rem}(c, d) \geq 0$, so $r > 0$ and so $d' > 0$. Finally,

$$\begin{aligned} \gcd(c', d') &= \gcd(d, r) && \text{by line 8} \\ &= \gcd(d, \text{rem}(c, d)) && \text{by line 4} \\ &= \gcd(c, d) && \text{see problem 1.18} \\ &= \gcd(m, n). && \text{by induction hypothesis} \end{aligned}$$

For partial correctness it is enough to show that if the algorithm terminates, the post-condition holds. If the algorithm terminates, then $r = 0$, so $\text{rem}(c, d) = 0$ and $\gcd(c, d) = \gcd(d, 0) = d$. On the other hand, by (LI), we have that $am + bn = d$, so $am + bn = d = \gcd(c, d)$ and $\gcd(c, d) = \gcd(m, n)$.

(c) On pp. 292–293 in [Delfs and Knebl (2007)] there is a nice analysis of their version of the algorithm. They bound the running time in terms of Fibonacci numbers, and obtain the desired bound on the running time.

Problem 1.23. For partial correctness of algorithm 1.3, we show that if the pre-condition holds, and *if* the algorithm terminates, then the post-condition will hold. So assume the pre-condition, and suppose first that A is *not* a palindrome. Then there exists a smallest i_0 (there exists one, and so by the LNP there exists a smallest one) such that $A[i_0] \neq A[n - i_0 + 1]$, and so, after the first $i_0 - 1$ iteration of the while-loop, we know from the loop invariant that $i = (i_0 - 1) + 1 = i_0$, and so line 4 is executed and the algorithm returns F. Therefore, "A not a palindrome" $\Rightarrow$ "return F."

Suppose now that A *is* a palindrome. Then line 4 is never executed (as no such i_0 exists), and so after the $k = \lfloor \frac{n}{2} \rfloor$-th iteration of the while-loop, we know from the loop invariant that $i = \lfloor \frac{n}{2} \rfloor + 1$ and so the while-loop is not executed any more, and the algorithm moves on to line 8, and returns T. Therefore, "A is a palindrome" $\Rightarrow$ "return T."

Therefore, the post-condition, "return T iff A is a palindrome," holds. Note that we have only used part of the loop invariant, that is we used the fact that after the k-th iteration, $i = k + 1$; it still holds that after the k-th iteration, for $1 \leq j \leq k$, $A[j] = A[n - j + 1]$, but we do not need this fact in the above proof.

To show that the algorithm does actually terminates, let $d_i = \lfloor \frac{n}{2} \rfloor - i$. By the pre-condition, we know that $n \geq 1$. The sequence $d_1, d_2, d_3, \ldots$ is a decreasing sequence of positive integers (because $i \leq \lfloor \frac{n}{2} \rfloor$), so by the LNP it is finite, and so the loop terminates.

Problem 1.24. It is very easy once you realize that in Python the slice `[::-1]` generates the reverse string. So, to check whether string `s` is a palindrome, all we do is write `s == s[::-1]`.

Problem 1.25. The solution is given by algorithm 1.16.

Let the loop invariant be: "x is a power of 2 iff n is a power of 2."

We show the loop invariant by induction on the number of iterations of the main loop. Basis case: zero iterations, and since $x \leftarrow n$, $x = n$, so obviously x is a power of 2 iff n is a power of 2. For the induction step, note that if we ever get to update x, we have $x' = x/2$, and clearly x' is a

Algorithm 1.16 Powers of 2.

Pre-condition: $n \geq 1$

```
x ← n
while (x > 1) do
        if (2|x) then
                x ← x/2
        else
                stop and return "no"
        end if
end while
return "yes"
```

Post-condition: "yes" $\iff$ n is a power of 2

power of 2 iff x is. Note that the algorithm always terminates (let $x_0 = n$, and $x_{i+1} = x_i/2$, and apply the LNP as usual).

We can now prove correctness: if the algorithms returns "yes", then after the last iteration of the loop $x = 1 = 2^0$, and by the loop invariant n is a power of 2. If, on the other hand, n is a power of 2, then so is every x, so eventually $x = 1$, and so the algorithm returns "yes".

Problem 1.26. Algorithm 1.4 computes the product of m and n, that is, the returned $z = m \cdot n$. A good loop invariant is $x \cdot y + z = m \cdot n$.

Problem 1.28. We are going to prove a loop invariant on the outer loop of algorithm 1.6, that is, we are going to prove a loop invariant on the for-loop (indexed on i) that starts on line 2 and ends on line 7. Our invariant consists of two parts: after the k-th iteration of the loop, the following two statements hold true:

(1) the set $\{v_1^*, \ldots, v_{k+1}^*\}$ is orthogonal, and
(2) $\text{span}\{v_1, \ldots, v_{k+1}\} = \text{span}\{v_1^*, \ldots, v_{k+1}^*\}$.

Basis case: after zero iterations of the for-loop, that is, before the for-loop is ever executed, we have, from line 1 of the algorithm, that $v_1^* \longleftarrow v_1$, and so the first statement is true because $\{v_1^*\}$ is orthogonal (a set consisting of a single non-zero vector is always orthogonal—and $v_1^* = v_1 \neq 0$ because the assumption (i.e., pre-condition) is that $\{v_1, \ldots, v_n\}$ is linearly independent, and so none of these vectors can be zero), and the second statement also holds trivially since if $v_1^* = v_1$ then $\text{span}\{v_1\} = \text{span}\{v_1^*\}$.

Induction Step: Suppose that the two conditions hold after the first k iterations of the loop; we are going to show that they continue to hold after the $k+1$ iteration. Consider:

$$v_{k+2}^{*} = v_{k+2} - \sum_{j=1}^{k+1} \mu_{(k+1)j} v_j^{*},$$

which we obtain directly from line 6 of the algorithm; note that the outer for-loop is indexed on i which goes from 2 to n, so after the k-th execution of line 2, for $k \geq 1$, the value of the index i is $k+1$. We show the first statement, i.e., that $\{v_1^{*}, \ldots, v_{k+2}^{*}\}$ are orthogonal. Since, by induction hypothesis, we know that $\{v_1^{*}, \ldots, v_{k+1}^{*}\}$ are already orthogonal, it is enough to show that for $1 \leq l \leq k+1$, $v_l^{*} \cdot v_{k+2}^{*} = 0$, which we do next:

$$\begin{aligned} v_l^{*} \cdot v_{k+2}^{*} &= v_l^{*} \cdot \left(v_{k+2} - \sum_{j=1}^{k+1} \mu_{(k+2)j} v_j^{*} \right) \\ &= (v_l^{*} \cdot v_{k+2}) - \sum_{j=1}^{k+1} \mu_{(k+2)j} (v_l^{*} \cdot v_j^{*}) \end{aligned}$$

and since $v_l^{*} \cdot v_j^{*} = 0$ unless $l = j$, we have:

$$= (v_l^{*} \cdot v_{k+2}) - \mu_{(k+2)l} (v_l^{*} \cdot v_l^{*})$$

and using line 4 of the algorithm we write:

$$= (v_l^{*} \cdot v_{k+2}) - \frac{v_{k+2} \cdot v_l^{*}}{\|v_l^{*}\|^2} (v_l^{*} \cdot v_l^{*}) = 0$$

where we have used the fact that $v_l \cdot v_l = \|v_l\|^2$ and that $v_l^{*} \cdot v_{k+2} = v_{k+2} \cdot v_l^{*}$.

For the second statement of the loop invariant we need to show that

$$\operatorname{span}\{v_1, \ldots, v_{k+2}\} = \operatorname{span}\{v_1^{*}, \ldots, v_{k+2}^{*}\}, \tag{1.7}$$

assuming, by the induction hypothesis, that $\operatorname{span}\{v_1, \ldots, v_{k+1}\} = \operatorname{span}\{v_1^{*}, \ldots, v_{k+1}^{*}\}$. The argument will be based on line 6 of the algorithm, which provides us with the following equality:

$$v_{k+2}^{*} = v_{k+2} - \sum_{j=1}^{k+1} \mu_{(k+2)j} v_j^{*}. \tag{1.8}$$

Given the induction hypothesis, to show (1.7) we need only show the following two things:

(1) $v_{k+2} \in \operatorname{span}\{v_1^{*}, \ldots, v_{k+2}^{*}\}$, and
(2) $v_{k+2}^{*} \in \operatorname{span}\{v_1, \ldots, v_{k+2}\}$.

Using (1.8) we obtain immediately that $v_{k+2} = v^*_{k+2} + \sum_{j=1}^{k+1} \mu_{(k+2)j} v^*_j$ and so we have (1). To show (2) we note that

$$\text{span}\{v_1, \ldots, v_{k+2}\} = \text{span}\{v^*_1, \ldots, v^*_{k+1}, v_{k+2}\}$$

by the induction hypothesis, and so we have what we need directly from (1.8).

Finally, note that we never divide by zero in line 4 of the algorithm because we always divide by $\|v^*_j\|$, and the only way for the norm to be zero is if the vector itself, v^*_j, is zero. But we know from the post-condition that $\{v^*_1, \ldots, v^*_n\}$ is a basis, and so these vectors must be linearly independent, and so none of them can be zero.

Problem 1.30. A reference for this algorithm can be found in [Hoffstein *et al.* (2008)] in §6.12.1. Also [von zur Gathen and Gerhard (1999)], §16.2, gives a treatment of the algorithm in higher dimensions.

Let $p = v_1 \cdot v_2/\|v_1\|^2$, and keep the following relationship in mind:

$$\lfloor p \rceil = \lfloor p + \frac{1}{2} \rfloor = m \in \mathbb{Z} \iff p \in [m - \frac{1}{2}, m + \frac{1}{2}) \subseteq \mathbb{R},$$

where, following standard calculus terminology, the set $[a, b)$, for $a, b \in \mathbb{R}$, denotes the set of all $x \in \mathbb{R}$ such that $a \le x < b$.

We now give a proof of termination. Suppose first that $|p| = \frac{1}{2}$. If $p = -\frac{1}{2}$, then $m = 0$ and the algorithm stops. If $p = \frac{1}{2}$, then $m = 1$, which means that we go through the loop one more time with $v'_1 = v_1$ and $\|v'_2\| = \|v_2 - v_1\| = \|v_2\|$, and, more importantly, in the next round $p = -\frac{1}{2}$, and again the algorithm terminates.

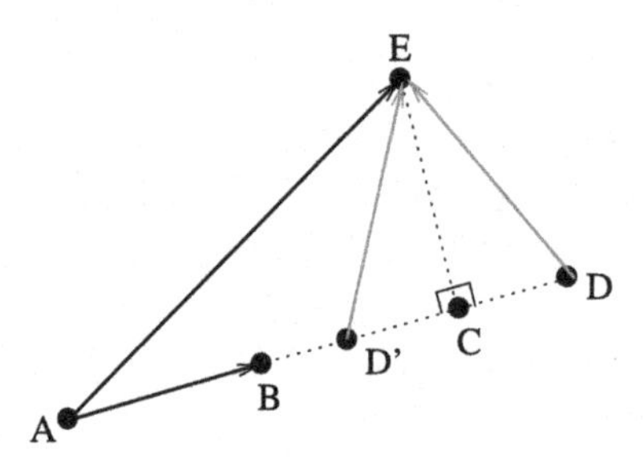

Fig. 1.7 The projection of v_2, given as $\vec{AE}$, onto v_1, given as $\vec{AB}$. The resulting vector is $\vec{AC} = v_2 - pv_1$, where $p = v_1 \cdot v_2/\|v_1\|^2$. Letting $m = \lfloor p \rceil$, the vector $v_2 - mv_1$, is given by $\vec{D'E}$ or $\vec{DE}$, depending on whether $m < p$ or not, respectively. Of course, $D' = C = D$ when $p = m$.

If $p = m$, i.e., p was an integer to begin with (giving $\vec{CE} = \vec{D'E} = \vec{DE}$ in figure 1.7), then simply by the Pythagorean theorem $\|\vec{CE}\|$ has to be shorter than $\|\vec{AE}\|$ (as v_1, v_2 are non-zero, as $m \neq 0$).

So we may assume that $|p| \neq \frac{1}{2}$ and $p \neq m$. The two cases where $m < p$, giving D', or $m > p$, giving D, are symmetric, and so we treat only the latter case. It must be that $|p| > \frac{1}{2}$ for otherwise m would have been zero, resulting in termination. Note that $\|\vec{CD}\| \leq \frac{1}{2}\|\vec{AB}\|$, because $\vec{AD} = m\vec{AB}$. From this and the Pythagorean theorem we know that:

$$\|\vec{AE}\|^2 = \|\vec{AC}\|^2 + \|\vec{CE}\|^2 = p^2\|\vec{AB}\|^2 + \|\vec{CE}\|^2$$

$$\|\vec{DE}\|^2 = \|\vec{CD}\|^2 + \|\vec{CE}\|^2 \leq p^2\|\vec{AB}\|^2 + \|\vec{CE}\|^2$$

and so $\|\vec{AE}\|^2 - \|\vec{DE}\|^2 \geq (p^2 - \frac{1}{4})\|\vec{AB}\|^2$, and, as we already noted, if the algorithm does not end in line 6 that means that $|p| > \frac{1}{2}$, and so it follows that $\|\vec{AE}\| > \|\vec{DE}\|$, that is, v_2 is longer than $v_2 - mv_1$, and so the new v_2 (line 9) is shorter than the old one.

Let v_1', v_2' be the two vectors resulting in one iteration of the loop from v_1, v_2. As we noted above, when $|p| = \frac{1}{2}$ termination comes in one or two steps. Otherwise, $\|v_1'\| + \|v_2'\| < \|v_1\| + \|v_2\|$, and as there are finitely many pairs of points in a lattice bounded by the sum of the two norms of the original vectors, and the algorithm ends when one of the vectors becomes zero, this procedure must end in finitely many steps.

Problem 1.32. Let $S \subseteq \mathbb{Z} \times \mathbb{Z}$ be the set consisting precisely of those pairs of integers (x, y) such that $x \geq y$ and $x - y$ is even. We are going to prove that S is the domain of definition of F. First, if $x < y$ then $x \neq y$ and so we go on to compute $F(x, F(x-1, y+1))$, and now we must compute $F(x-1, y+1)$; but if $x < y$, then clearly $x - 1 < y + 1$; this condition is preserved, and so we end up having to compute $F(x-i, y+i)$ for all i, and so this recursion never "bottoms out." Suppose that $x - y$ is odd. Then $x \neq y$ (as 0 is even!), so again we go on to $F(x, F(x-1, y+1))$; if $x - y$ is odd, so is $(x-1) - (y+1) = x - y - 2$. Again we end up having to compute $F(x-i, y+i)$ for all i, and so the recursion never terminates. Clearly, all the pairs in S^c are not in the domain of definition of F.

Suppose now that $(x, y) \in S$. Then $x \geq y$ and $x - y$ is even; thus, $x - y = 2i$ for some $i \geq 0$. We show, by induction on i, that the algorithm terminates on such (x, y) and outputs $x + 1$. Basis case: $i = 0$, so $x = y$, and so the algorithm returns $y + 1$ which is $x + 1$. Suppose now that $x - y = 2(i+1)$. Then $x \neq y$, and so we compute $F(x, F(x-1, y+1))$. But

$$(x-1) - (y+1) = x - y - 2 = 2(i+1) - 2 = 2i,$$

for $i \geq 0$, and so by induction $F(x-1, y+1)$ terminates and outputs $(x-1)+1 = x$. So now we must compute $F(x, x)$ which is just $x+1$, and we are done.

Problem 1.33. We show that f_1 is a fixed point of algorithm 1.9. Recall that in problem 1.32 we showed that the domain of definition of F, the function computed by algorithm 1.9, is $S = \{(x, y) : x - y = 2i, i \geq 0\}$. Now we show that if we replace F in algorithm 1.9 by f_1, the new algorithm, which is algorithm 1.17, still computes F albeit not recursively (as f_1 is defined by algorithm 1.10 which is not recursive).

Algorithm 1.17 Algorithm 1.9 with F replaced by f_1.

1: **if** $x = y$ **then**
2:     **return** $y + 1$
3: **else**
4:     $f_1(x, f_1(x-1, y+1))$
5: **end if**

We proceed as follows: if $(x, y) \in S$, then $x - y = 2i$ with $i \geq 0$. On such (x, y) we know, from problem 1.32, that $F(x, y) = x+1$. Now consider the output of algorithm 1.17 on such a pair (x, y). If $i = 0$, then it returns $y + 1 = x + 1$, so we are done. If $i > 0$, then it computes

$$f_1(x, f_1(x-1, y+1)) = f_1(x, x) = x + 1,$$

and we are done. To see why $f_1(x-1, y+1) = x$ notice that there are two cases: first, if $x - 1 = y + 1$, then the algorithm for f_1 (algorithm 1.10) returns $(y+1)+1 = (x-1)+1 = x$. Second, if $x - 1 > y + 1$ (and that is the only other possibility), algorithm 1.10 returns $(x-1)+1 = x$ as well.

Problem 1.36. After b proposed to g for the first time, whether this proposal was successful or not, the partners of g could have only gotten better. Thus, there is no need for b to try again.

Problem 1.37. b_{s+1} proposes to the girls according to his list of preference; a g ends up accepting, and if the g who accepted b_{s+1} was free, she is the new one with a partner. Otherwise, some $b^* \in \{b_1, \ldots, b_s\}$ became disengaged, and we repeat the same argument. The g's disengage only if a better b proposes, so it is true that $p_{M_{s+1}}(g_j) <^j p_{M_s}(g_j)$.

Problem 1.38. Suppose that we have a blocking pair $\{b, g\}$ (meaning that $\{(b, g'), (b', g)\} \subseteq M_n$, but b prefers g to g', and g prefers b to b'). Either b came after b' or before. If b came before b', then g would have been with b or someone better when b' came around, so g would not have become engaged to b'. On the other hand, since (b', g) is a pair, no better offer has been made to g after the offer of b', so b could not have come after b'. In either case we get an impossibility, and so there is no blocking pair $\{b, g\}$.

Problem 1.39. To show that the matching is boy-optimal, we argue by contradiction. Let "*g is an optimal partner for b*" mean that among all the stable matchings g is the best partner that b can get.

We run the Gale-Shapley algorithm, and let b be the first boy who is rejected by his optimal partner g. This means that g has already been paired with some b', and g prefers b' to b. Furthermore, g is at least as desirable to b' as his own optimal partner (since the proposal of b is the first time during the run of the algorithm that a boy is rejected by his optimal partner). Since g is optimal for b, we know (by definition) that there exists some stable matching S where (b, g) is a pair. On the other hand, the optimal partner of b' is ranked (by b' of course) at most as high as g, and since g is taken by b, whoever b' is paired with in S, say g', b' prefers g to g'. This gives us an unstable pairing, because $\{b', g\}$ prefer each other to the partners they have in S.

To show that the Gale-Shapley algorithm is girl-pessimal, we use the fact that it is boy-optimal (which we just showed). Again, we argue by contradiction. Suppose there is a stable matching S where g is paired with b, and g prefers b' to b, where (b', g) is the result of the Gale-Shapley algorithm. By boy-optimality, we know that in S we have (b', g'), where g' is not higher on the preference list of b' than g, and since g is already paired with b, we know that g' is actually lower. This says that S is unstable since $\{b', g\}$ would rather be together than with their partners.

1.6 Notes

This book is about proving things about algorithms; their correctness, their termination, their running time, etc. The art of mathematical proofs is a difficult art to master; a very good place to start is [Velleman (2006)].

$\mathbb{N}$ (the set of natural numbers) and IP (the induction principle) are very tightly related; the rigorous definition of $\mathbb{N}$, as a set-theoretic object, is the following: it is the *unique* set satisfying the following three properties: (i) it contains 0, (ii) if n is in it, then so is $n+1$, and (iii) it satisfies the induction principle (which in this context is stated as follows: if S is a subset of $\mathbb{N}$, and S satisfies (i) and (ii) above, then in fact $S = \mathbb{N}$).

The references in this paragraph are with respect to Knuth's seminal *The Art of Computer Programming*, [Knuth (1997)]. For an extensive study of Euclid's algorithm see §1.1. Problem 1.2 comes from §1.2.1, problem #8, pg. 19. See §2.3.1, pg. 318 for more background on tree traversals. For the history of the concept of pre and post-condition, and loop invariants, see pg. 17. In particular, for material related to the extended Euclid's algorithm , see page 13, algorithm E, in [Knuth (1997)], page 937 in [Cormen *et al.* (2009)], and page 292, algorithm A.5, in [Delfs and Knebl (2007)]. We give a recursive version of the algorithm in section 3.4.

See [Zingaro (2008)] for a book dedicated to the idea of invariants in the context of proving correctness of algorithms. A great source of problems on the invariance principle, that is section 1.2, is chapter 1 in [Engel (1998)]

The example about the 8×8 board with two squares missing (figure 1.2) comes from [Dijkstra (1989)].

The palindrome `madamimadam` comes from Joyce's *Ulysses.*

Section 1.3.5 on the correctness of recursive algorithms is based on chapter 5 of [Manna (1974)].

Section 1.4 is based on §2 in [Cenzer and Remmel (2001)]. For another presentation of the Stable Marriage problem see chapter 1 in [Kleinberg and Tardos (2006)]. The reference to the Marquis de Condorcet in the first sentence of section 1.4 comes from the PhD thesis of Yun Zhai ([Zhai (2010)]), written under the supervision of Ryszard Janicki. In that thesis, Yun Zhai references [Arrow (1951)] as the source of the remark regarding the Marquis de Condorcet's early attempts at pairwise ranking.

Chapter 2

Greedy Algorithms

Greedy algorithms are algorithms prone to instant gratification. Without looking too far ahead, at each step they make a *locally optimum* choice, with the hope that it will lead to a *global optimum* at the end.

An example of a greedy procedure is a convenience store clerk dispensing change. In order to use as few coins as possible, the clerk gives out the coins of highest value for as long as possible, before moving on to a smaller denomination.

Being greedy is a simple heuristic that works well with some computational problems but fails with others. In the case of cash dispensing, if we have coins of value $1, 5, 25$ the greedy procedure always produces the smallest possible (i.e., optimal) number of coins. But if we have coins of value $1, 10, 25$, then dispensing 30 greedily results in six coins $(25, 1, 1, 1, 1, 1)$, while a more astute—and less greedy—clerk would give out only three coins $(10, 10, 10)$. See section 2.3.1 for the make-change problem.

2.1 Minimum cost spanning trees

We start by giving several graph-theoretic definitions. Note that our graphs are assumed to be finite (i.e., they have finitely many nodes), and represented using an *adjacency matrix*. Given a directed or undirected graph $G = (V, E)$—all the definitions are given below—its adjacency matrix is a matrix A_G of size $n \times n$, where $n = |V|$, such that entry (i, j) is 1 if (i, j) is an edge in G, and it is 0 otherwise.

An adjacency matrix itself can be easily encoded as a string over $\{0, 1\}$. That is, given A_G of size $n \times n$, let $s_G \in \{0, 1\}^{n^2}$, where s_G is simply the concatenation of the rows of A_G. We can check directly from s_G if (i, j) is an edge by checking if position $(i - 1)n + j$ in s_G contains a 1.

An *undirected graph* G is a pair (V, E) where V is a set (of vertices or nodes), and $E \subseteq V \times V$ and $(u, v) \in E$ iff $(v, u) \in E$, and $(u, u) \notin E$. The *degree* of a vertex v is the number of edges touching v. A *path* in G between v_1 and v_k is a sequence $v_1, v_2, \ldots, v_k$ such that each $(v_i, v_{i+1}) \in E$. G is *connected* if between every pair of distinct nodes there is a path. A *cycle* is a simply closed path $v_1, \ldots, v_k, v_1$ with $v_1, \ldots, v_k$ all distinct, and $k \geq 3$. A graph is *acyclic* if it has no cycles. A *tree* is a connected acyclic graph. A *spanning tree* of a connected graph G is a subset $T \subseteq E$ of the edges such that (V, T) is a tree. In other words, the edges in T must connect all nodes of G and contain no cycles.

If G has a cycle, then there is more than one spanning tree for G, and in general G may have many spanning trees, but each spanning tree has the same number of edges.

Lemma 2.1. *Every tree with n nodes has exactly $n - 1$ edges.*

Problem 2.2. *Prove lemma 2.1. (Hint: first show that every tree has a* leaf, *i.e., a node of degree one. Then show the lemma by induction on n.)*

Lemma 2.3. *A graph with n nodes and more than $n-1$ edges must contain at least one cycle.*

Problem 2.4. *Prove lemma 2.3.*

It follows from lemmas 2.1 and 2.3 that if a graph is a tree, i.e., it is acyclic and connected, then it must have $(n - 1)$ edges. If it does *not* have $(n - 1)$ edges, then it is either not acyclic, or it is not connected. If it has less than $(n - 1)$ edges, it is certainly not connected, and if it has more than $(n - 1)$ edges, it is certainly not acyclic.

We are interested in finding a minimum cost spanning tree for G, assuming that each edge e is assigned a cost $c(e)$. The understanding is that the costs are non-negative real number, i.e., each $c(e)$ is in $\mathbb{R}^+$. The total cost $c(T)$ is the sum of the costs of the edges in T. We say that T is a *minimum cost spanning tree (MCST)* for G if T is a spanning tree for G and given any spanning tree T' for G, $c(T) \leq c(T')$.

Given a graph $G = (V, E)$, where $c(e_i)$ = "cost of edge e_i," we want to find a MCST. It turns out, fortuitously, that an obvious greedy algorithm—known as Kruskal's algorithm—works. The algorithm is: sort the edges in non-decreasing order of costs, so that $c(e_1) \leq c(e_2) \leq \ldots \leq c(e_m)$, and add the edges one at a time, except when including an edge would form a cycle with the edges added already.

Algorithm 2.1 Kruskal

```
1: Sort the edges: c(e_1) ≤ c(e_2) ≤ ... ≤ c(e_m)
2: T ⟵ ∅
3: for i : 1..m do
4:     if T ∪ {e_i} has no cycle then
5:         T ⟵ T ∪ {e_i}
6:     end if
7: end for
```

But how do we test for a cycle, i.e., execute line 4 in algorithm 2.1? At the end of each iteration of the for-loop, the set T of edges divides the vertices V into a collection $V_1, \ldots, V_k$ of *connected components*. That is, V is the disjoint union of $V_1, \ldots, V_k$, each V_i forms a connected graph using edges from T, and no edge in T connects V_i and V_j, if $i \neq j$. A simple way to keep track of $V_1, \ldots, V_k$ is to use an array $D[i]$ where $D[i] = j$ if vertex $i \in V_j$. Initialize D by setting $D[i] \longleftarrow i$ for every $i = 1, 2, \ldots, n$.

To check whether $e_i = (r, s)$ forms a cycle within T, it is enough to check whether $D[r] = D[s]$. If e_i does not form a cycle within T, then we update: $T \longleftarrow T \cup \{(r, s)\}$, and we merge the component $D[r]$ with $D[s]$ as shown in algorithm 2.2.

Algorithm 2.2 Merging components

```
k ⟵ D[r]
l ⟵ D[s]
for j : 1..n do
    if D[j] = l then
        D[j] ⟵ k
    end if
end for
```

Problem 2.5. *Given that the edges can be ordered in m^2 steps, with, for example, insertion sort, what is the running time of algorithm 2.1?*

Problem 2.6. *Write a Python program that implements Kruskal's algorithm (algorithm 2.1) using algorithm 2.2 for merging components. You will need a data structure in order to represent the graph—on page 39 we explain what is an adjacency matrix for a graph; use that, with entry (i, j) being $c(i, j)$, the cost of the edge (i, j), and zero if there is no such edge.*

You may also choose -1 *to represent "no edge" if you want to have edges of zero cost. Note that for an undirected graph, the adjacency matrix is symmetric; thus, in order to save space, you may want to keep track only of the entries above the main diagonal. The input to your program should be an adjacency matrix, and the output should also be an adjacency matrix.*

We now prove that Kruskal's algorithm works. It is not immediately clear that Kruskal's algorithm yields a spanning tree, let alone a MCST. To see that the resulting collection T of edges is a spanning tree for G (assuming that G is connected), we must show that (V, T) is connected and acyclic.

It is obvious that T is acyclic, because we never add an edge that results in a cycle. To show that (V, T) is connected, we reason as follows. Let u and v be two distinct nodes in V. Since G is connected, there is a path p connecting u and v in G. The algorithm considers each edge e_i of G in turn, and puts e_i in T *unless* $T \cup \{e_i\}$ forms a cycle. But in the latter case, there must already be a path in T connecting the end points of e_i, so deleting e_i does not disconnect the graph.

This argument can be formalized by showing that the following statement is an invariant of the loop in Kruskal's algorithm:

$$\textit{The edge set } T \cup \{e_{i+1}, \ldots, e_m\} \textit{ connects all nodes in } V. \tag{2.1}$$

Problem 2.7. *Prove, by induction, that (2.1) is a loop invariant. First show that (2.1) holds before the main loop of Kruskal's algorithm executes (the* 0*-th iteration; this is the basis case—remember the assumption (*pre-condition*) that* $G = (V, E)$ *is connected). Then show that if (2.1) holds after the* i*-th execution of the loop, then* $T \cup \{e_{i+2}, \ldots, e_m\}$ *connects all nodes of* V *after the* $(i+1)$*-st execution of the loop. Conclude by induction that (2.1) holds for all* i*. Finally, show how to use this loop invariant to prove that* T *is connected.*

Problem 2.8. *Suppose that* $G = (V, E)$ *is* not *connected. Show that in this case, when* G *is given to Kruskal's algorithm as input, the algorithm computes a* spanning forest *of* G*. Define first precisely what is a spanning forest (first define the notion of a* connected component*). Then give a formal proof using the idea of a loop invariant, as in problem 2.7.*

To show that the spanning tree resulting from the algorithm is in fact a MCST, we reason that after each iteration of the loop, the set T of edges can be extended to a MCST using edges that have *not yet* been considered.

Hence after termination, all edges have been considered, so T must itself be a MCST. We say that a set T of edges of G is *promising* if T can be extended to a MCST for G, that is, T is promising if there exists a MCST T' such that $T \subseteq T'$.

Lemma 2.9. *"T is promising" is a loop invariant for Kruskal's algorithm.*

Proof. The proof is by induction on the number of iterations of the main loop of Kruskal's algorithm. Basis case: at this stage the algorithm has gone through the loop zero times, and initially T is the empty set, which is obviously promising (the empty set is a subset of any set).

Induction step: We assume that T is promising, and show that T continues being promising after one more iteration of the loop.

Notice that the edges used to expand T to a spanning tree must come from those not yet considered, because the edges that have been considered are either in T already, or have been rejected because they form a cycle. We examine by cases what happens after edge e_i has been considered:
Case 1: e_i is rejected. T remains unchanged, and it is still promising. There is one subtle point: T was promising before the loop was executed, meaning that there was a subset of edges $S \subseteq \{e_i, \ldots, e_m\}$ that extended T to a MCST, i.e., $T \cup S$ is a MCST. But after the loop is executed, the edges extending T to a MCST would come from $\{e_{i+1}, \ldots, e_m\}$; but this is not a problem, as e_i could not be part of S (as then $T \cup S$ would contain a cycle), so $S \subseteq \{e_{i+1}, \ldots, e_m\}$, and so S is still a candidate for extending T to a MCST, even *after* the execution of the loop. Thus T remains promising after the execution of the loop, though the edges extending it to a MCST come from a smaller set (i.e., not containing e_i).
Case 2: e_i is accepted. We must show that $T \cup \{e_i\}$ is still promising. Since T is promising, there is a MCST T_1 such that $T \subseteq T_1$. We consider two subcases.
Subcase a: $e_i \in T_1$. Then obviously $T \cup \{e_i\}$ is promising.
Subcase b: $e_i \notin T_1$. Then, according to the Exchange Lemma below, there is an edge e_j in $T_1 - T_2$, where T_2 is the spanning tree resulting from the algorithm, such that $T_3 = (T_1 \cup \{e_i\}) - \{e_j\}$ is a spanning tree. Notice that $i < j$, since otherwise e_j would have been rejected from T and thus would form a cycle in T and so also in T_1. Therefore $c(e_i) \leq c(e_j)$, so $c(T_3) \leq c(T_1)$, so T_3 must also be a MCST. Since $T \cup \{e_i\} \subseteq T_3$, it follows that $T \cup \{e_i\}$ is promising.

This finishes the proof of the induction step. □

Lemma 2.10 (Exchange Lemma). *Let G be a connected graph, and let T_1 and T_2 be any two spanning trees for G. For every edge e in $T_2 - T_1$ there is an edge e' in $T_1 - T_2$ such that $T_1 \cup \{e\} - \{e'\}$ is a spanning tree for G. (See figure 2.1.)*

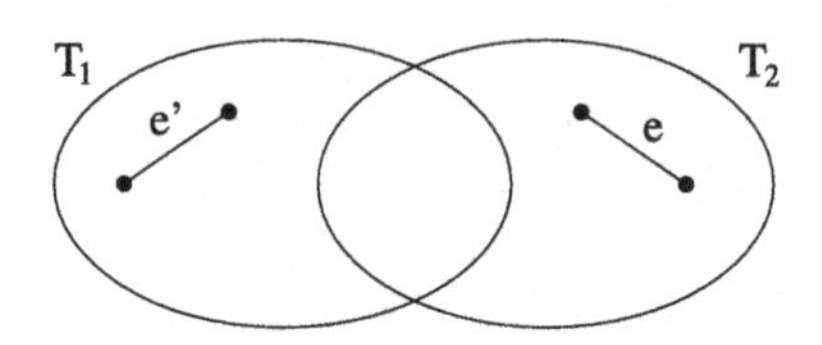

Fig. 2.1 Exchange lemma.

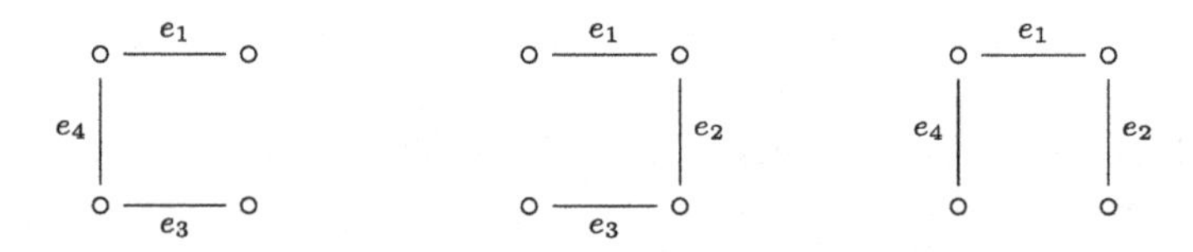

Fig. 2.2 Example of the exchange lemma: the left-most and the middle graphs are two different spanning trees of the same graph. Suppose we add edge e_4 to the middle tree; then we delete e_3 and obtain the right-most spanning tree.

Problem 2.11. *Prove this lemma. (Hint: let e be an edge in $T_2 - T_1$. Then $T_1 \cup \{e\}$ contains a cycle—can all the edges in this cycle belong to T_2?).*

Problem 2.12. *Suppose that edge e_1 has a smaller cost than any of the other edges; that is, $c(e_1) < c(e_i)$, for all $i > 1$. Show that there is at least one MCST for G that includes edge e_1.*

Problem 2.13. *Before algorithm 2.1 proceeds, it orders the edges in line 1, and presumably breaks ties—i.e., sorts edges of the same cost—arbitrarily. Show that for every MCST T of a graph G, there exists a particular way of breaking the ties so that the algorithm returns T.*

Problem 2.14. *Write a Python program that takes as input the description of a grid, and outputs its minimum cost spanning tree. An n-grid is a graph consisting of n^2 nodes, organized as a square array of $n \times n$ points. Every node may be connected to at most the nodes directly above and below (if*

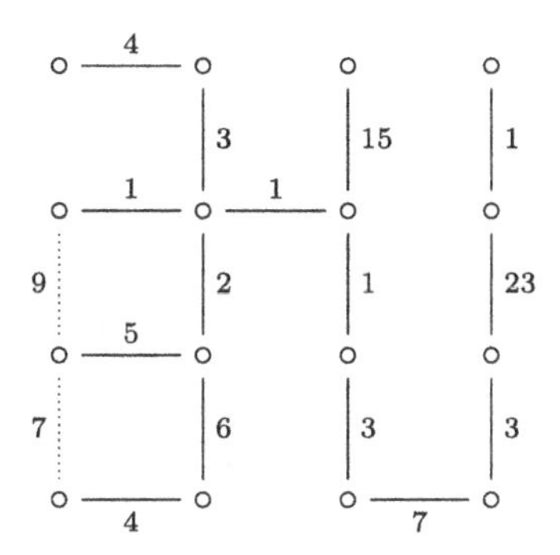

Fig. 2.3 An example of a 4-grid. Note that it has $4^2 = 16$ nodes, and 17 edges.

they exist), and to the two nodes immediately to the left and right (if they exist). An example of a 4-grid is given in figure 2.3.

What is the largest number of edges that an n-grid may have? We have the following node-naming convention: we name the nodes from left-to-right, row-by-row, starting with the top row. Thus, our 4-grid is described by the following adjacency list:

$$4 : (1, 2; 4), (2, 6; 3), (3, 7; 15), (4, 8; 1), (5, 6; 1), (6, 7; 1), \ldots \tag{2.2}$$

where the first integer is the grid size parameter, and the first two integers in each triple denote the two (distinct) nodes that describe (uniquely) an edge, and the third integer, following the semicolon, gives the cost of that edge.

When given as input a list of triples—such as (2.2)—your program must first check whether the list describes a grid, and then compute the minimum cost spanning tree of the grid. In our 4-grid example, the solid edges describe a minimum cost spanning tree. Also note that the edges in (2.2) are not required to be given in any particular order.

Your program should take as input a file, say `graph.txt`*, containing a list such as (2.1). For example,* `2:(1,2;9),(3,4;5),(2,4;6),(1,3;2)` *and it should output, directly to the screen, a graph indicating the edges of a minimum cost spanning tree. The graph should be "text-based" with "*" describing nodes and "-" and "|" describing edges. In this example, the MCST of the given 2-grid would be represented as:*

```
* *
| |
*-*
```

2.2 Jobs with deadlines and profits

We have n jobs, each of which takes unit time, and a processor on which we would like to schedule them sequentially in as profitable a manner as possible. Each job has a profit associated with it, as well as a deadline; if a job is not scheduled by its deadline, then we do not get its profit. Because each job takes the same amount of time, we think of a schedule S as consisting of a sequence of job "slots" $1, 2, 3, \ldots$, where $S(t)$ is the job scheduled in slot t.

Formally, the input is a sequence of pairs $(d_1, g_1), (d_2, g_2), \ldots, (d_n, g_n)$ where $g_i \in \mathbb{R}^+$ is the profit obtainable from job i, and $d_i \in \mathbb{N}$ is the deadline for job i. In section 4.5 we are going to consider the case where jobs have arbitrary durations—given by a positive integer. However, when durations are arbitrary, rather than of the same unit value, a greedy approach does not "seem"[1] to work.

A *schedule* is an array $S(1), S(2), \ldots, S(d)$ where $d = \max d_i$, that is, d is the latest deadline, beyond which no jobs can be scheduled. If $S(t) = i$, then job i is scheduled at time t, $1 \leq t \leq d$. If $S(t) = 0$, then no job is scheduled at time t. A schedule S is *feasible* if it satisfies two conditions:
Condition 1: If $S(t) = i > 0$, then $t \leq d_i$, i.e., every scheduled job meets its deadline.
Condition 2: If $t_1 \neq t_2$ and also $S(t_1) \neq 0$, then $S(t_1) \neq S(t_2)$, i.e., each job is scheduled at most once.

Problem 2.15. *Write a Python program that takes as input a schedule S, and a sequence of jobs, and checks whether S is feasible.*

Let the total profit of schedule S be $P(S) = \sum_{t=1}^{d} g_{S(t)}$, where $g_0 = 0$. We want to find a feasible schedule S whose profit $P(S)$ is as large as possible; this can be accomplished with the greedy algorithm 2.3, which orders jobs in non-increasing order of profits and places them as late as possible within their deadline. It is surprising that this algorithm works, and it seems to be a scientific confirmation of the benefits of procrastination.

Note that line 7 in algorithm 2.3 finds the latest possible free slot that meets the deadline; if no such free slot exists, then job i cannot be scheduled. More precisely, the understanding is that if there is no t satisfying both $S(t) = 0$ and $t \leq d_i$, then the last command on line 7, $S(t) \longleftarrow i$, is *not* executed, and the for-loop considers the next i.

[1] We say "seem" in quotes because there is no known proof that a greedy algorithm will not do; such a proof would require a precise definition of what it means for a solution to be given by a greedy algorithm—a difficult task in itself.

Algorithm 2.3 Job scheduling

1: Sort the jobs in non-increasing order of profits: $g_1 \geq g_2 \geq \ldots \geq g_n$
2: $d \longleftarrow \max_i d_i$
3: **for** $t : 1..d$ **do**
4: $\quad S(t) \longleftarrow 0$
5: **end for**
6: **for** $i : 1..n$ **do**
7: $\quad$ Find the largest t such that $S(t) = 0$ and $t \leq d_i$, $S(t) \longleftarrow i$
8: **end for**

Problem 2.16. *Implement in Python algorithm 2.3 for job scheduling.*

Theorem 2.17. *The greedy solution to job scheduling is optimal, that is, the profit $P(S)$ of the schedule S computed by algorithm 2.3 is as large as possible.*

A schedule is *promising* if it can be extended to an optimal schedule. Schedule S' *extends* schedule S if for all $1 \leq t \leq d$, if $S(t) \neq 0$, then $S(t) = S'(t)$. For example, $S' = (2, 0, 1, 0, 3)$ extends $S = (2, 0, 0, 0, 3)$.

Lemma 2.18. *"S is promising" is an invariant for the (second) for-loop in algorithm 2.3.*

In fact, just as in the case of Kruskal's algorithm in the previous section, we must make the definition of "promising" in lemma 2.18 more precise: we say that "S is promising *after* the i-th iteration of the loop in algorithm 2.3" if S can be extended to an optimal schedule using jobs from those among $\{i+1, i+2, \ldots, n\}$, i.e., using a subset of those jobs that have not been considered yet.

Problem 2.19. *Consider the following input*

$$\{\underbrace{(1,10)}_{1}, \underbrace{(1,10)}_{2}, \underbrace{(2,8)}_{3}, \underbrace{(2,8)}_{4}, \underbrace{(4,6)}_{5}, \underbrace{(4,6)}_{6}, \underbrace{(4,6)}_{7}, \underbrace{(4,6)}_{8}\},$$

where the jobs have been numbered underneath for convenience. Trace the workings of algorithm 2.3 on this input. On the left place the job numbers in the appropriate slots; on the right, show how the optimal solution is adjusted to keep the "promising" property. Start in the following configuration:

$S^0 =$ | 0 | 0 | 0 | 0 | and $S^0_{\text{opt}} =$ | 2 | 4 | 5 | 8 |

$S =$		0		0		j		
$S_{\text{opt}} =$		0		i		j		

Fig. 2.4 If S has job j in a position, then S_{opt} has also job j in *the same* position. If S has a zero in a given position (no job is scheduled there) then S_{opt} may have zero or a different job in the same position.

Problem 2.20. *Why does lemma 2.18 imply theorem 2.17? (Hint: this is a simple observation).*

We now prove lemma 2.18.

Proof. The proof is by induction. Basis case: after the 0-th iteration of the loop, $S = (0, 0, \ldots, 0)$ and we may extend it with jobs $\{1, 2, \ldots, n\}$, i.e., we have all the jobs at our disposal; so S is promising, as we can take *any* optimal schedule, and it will be an extension of S.

Induction step: Suppose that S is promising, and let S_{opt} be *some* optimal schedule that extends S. Let S' be the result of one more iteration through the loop where job i is considered. We must prove that S' continues being promising, so the goal is to show that there is an optimal schedule S'_{opt} that extends S'. We consider two cases:
Case 1: job i cannot be scheduled. Then $S' = S$, so we let $S'_{\text{opt}} = S_{\text{opt}}$, and we are done. The only subtle thing is that S was extendable into S_{opt} with jobs in $\{i, i+1, \ldots, n\}$, but after the i-th iteration we no longer have job i at our disposal.

Problem 2.21. *Show that this "subtle thing" mentioned in the paragraph above is not a problem.*

Case 2: job i is scheduled at time t_0, so $S'(t_0) = i$ (whereas $S(t_0) = 0$) and t_0 is the latest possible time for job i in the schedule S. We have two subcases.
Subcase a: job i is scheduled in S_{opt} at time t_1:

If $t_1 = t_0$, then, as in case 1, just let $S'_{\text{opt}} = S_{\text{opt}}$.

If $t_1 < t_0$, then let S'_{opt} be S_{opt} except that we interchange t_0 and t_1, that is we let $S'_{\text{opt}}(t_0) = S_{\text{opt}}(t_1) = i$ and $S'_{\text{opt}}(t_1) = S_{\text{opt}}(t_0)$. Then S'_{opt} is feasible (why 1?), it extends S' (why 2?), and $P(S'_{\text{opt}}) = P(S_{\text{opt}})$ (why 3?).

The case $t_1 > t_0$ is not possible (why 4?).

Subcase b: job i is not scheduled in S_{opt}. Then we simply define S'_{opt} to be the same as S_{opt}, except $S'_{\text{opt}}(t_0) = i$. Since S_{opt} is feasible, so is S'_{opt}, and since S'_{opt} extends S', we only have to show that $P(S'_{\text{opt}}) = P(S_{\text{opt}})$. This follows from the following claim:

Claim 2.22. *Let $S_{\text{opt}}(t_0) = j$. Then $g_j \leq g_i$.*

Proof. We prove the claim by contradiction: assume that $g_j > g_i$ (note that in this case $j \neq 0$). Then job j was considered before job i. Since job i was scheduled at time t_0, job j must have been scheduled at time $t_2 \neq t_0$ (we know that job j was scheduled in S since $S(t_0) = 0$, and $t_0 \leq d_j$, so there was a slot for job j, and therefore it was scheduled). But S_{opt} extends S, and $S(t_2) = j \neq S_{\text{opt}}(t_2)$—contradiction. □

This finishes the proof of the induction step. □

Problem 2.23. *Make sure you can answer all the "why's" in the above proof. Also, where in the proof of the claim we use the fact that $j \neq 0$?*

Problem 2.24. *Under what condition on the inputs is there a unique optimal schedule? If there is more than one optimal schedule, and given one such optimal schedule, is there always an arrangement of the jobs, still in a non-decreasing order of profits, that results in the algorithm outputting this particular optimal schedule?*

2.3 Further examples and problems

2.3.1 *Make change*

The make-change problem, briefly described in the introduction to this chapter, consists in paying a given amount using the least number of coins, using some fixed denomination, and an unlimited supply of coins of each denomination.

Consider the following greedy algorithm to solve the make-change problem, where the denominations are $C = \{1, 10, 25, 100\}$. On input $n \in \mathbb{N}$, the algorithm outputs the smallest list L of coins (from among C) whose sum equals n.

Note that s equals the sum of the values of the coins in L, and that strictly speaking L is a *multiset* (the same element may appear more than once in a multiset).

Algorithm 2.4 Make change

1: $C \longleftarrow \{1, 10, 25, 100\}$; $L \longleftarrow \emptyset$; $s \longleftarrow 0$
2: **while** $(s < n)$ **do**
3: find the largest x in C such that $s + x \leq n$
4: $L \longleftarrow L \cup \{x\}$; $s \longleftarrow s + x$
5: **end while**
6: **return** L

Problem 2.25. *Implement in Python algorithm 2.4 for making change.*

Problem 2.26. *Show that algorithm 2.4 (with the given denominations) does* not *necessarily produce an optimal solution. That is, present an n for which the output L contains more coins than the optimal solution.*

Problem 2.27. *Suppose that $C = \{1, p, p^2, \ldots, p^n\}$, where $p > 1$ and $n \geq 0$ are integers. That is, "$C \longleftarrow \{1, 10, 25, 100\}$" in line 1 of algorithm 2.4 is replaced by "$C \longleftarrow \{1, p, p^2, \ldots, p^n\}$." Show that with this series of denominations (for some fixed p, n) the greedy algorithm above always finds an optimal solution. (Hint: Start with a suitable definition of a* promising list.*)*

2.3.2 *Maximum weight matching*

Let $G = (V_1 \cup V_2, E)$ be a bipartite graph, with edge set $E \subseteq V_1 \times V_2$, and $w : E \longrightarrow \mathbb{N}$ assigns a weight $w(e) \in \mathbb{N}$ to each edge $e \in E = \{e_1, \ldots, e_m\}$. A *matching* for G is a subset $M \subseteq E$ such that no two edges in M share a common vertex. The weight of M is $w(M) = \sum_{e \in M} w(e)$.

Problem 2.28. *Give a simple greedy algorithm which, given a bipartite graph with edge weights,* attempts *to find a matching with the largest possible weight.*

Problem 2.29. *Give an example of a bipartite graph with edge weights for which your algorithm in problem 2.28* fails *to find a matching with the largest possible weight.*

Problem 2.30. *Suppose all edge weights in the bipartite graph are distinct, and each is a power of 2. Prove that your greedy algorithm always succeeds in finding a maximum weight matching in this case. (Assume for this question that all the edges are there, i.e., that $E = V \times V$.)*

2.3.3 *Shortest path*

The following example of a greedy algorithm is very beautiful. It reminds one of the cartographers of old, who produced maps of the world with white spots—the unknown and unexplored places.

Suppose that we are given a graph $G = (V, E)$, a designated start node s, and a cost function for each edge $e \in E$, denoted $c(e)$. We are asked to compute the cheapest paths from s to every other node in G, where the cost of a path is the sum of the costs of its edges.

Consider the following greedy algorithm: the algorithm maintains a set S of explored nodes, and for each $u \in S$ it stores a value $d(u)$, which is the cheapest path inside S, starting at s and ending at u.

Initially, $S = \{s\}$ and $d(s) = 0$. Now, for each $v \in V - S$ we find the shortest path to v by traveling inside the explored part S to some $u \in S$, followed by a single edge (u, v). See figure 2.5.

That is, we compute:

$$d'(v) = \min_{u \in S, e=(u,v)} d(u) + c(e). \tag{2.3}$$

We choose the node $v \in V - S$ for which (2.3) is minimized, add v to S, and set $d(v) = d'(v)$, and repeat. Thus we add one node at a time to the explored part, and we stop when $S = V$.

This greedy algorithm for computing the shortest path is due to Edsger Dijkstra. It is not difficult to see that its running time is $O(n^2)$.

Problem 2.31. *Design the algorithm in pseudocode, and show that at the end, for each $u \in V$, $d(u)$ is the cost of the cheapest path from s to u.*

Problem 2.32. *The* Open Shortest Path First *(OSPF) is a routing protocol for IP, described in detail in* RFC 2328 *(where* RFC *stands for "Request for Comment," which is a series of memoranda published by the Internet Engineering Task Force describing the working of the Internet). The commonly used routing protocol OSPF uses Dijkstra's greedy algorithm for computing*

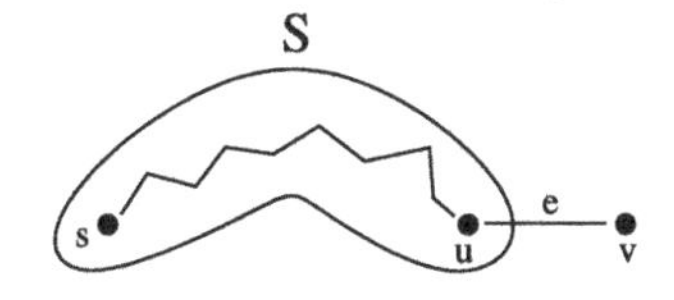

Fig. 2.5 Computing the shortest path.

the so called "shortest paths tree," which for a particular node x on the Internet, lists the best connections to all other nodes on x's subnetwork.

Write a Python program that implements a simplified dynamic routing policy mechanism. *More precisely, you are to implement a routing table management daemon, which maintains a link-state database as in the OSPF interior routing protocol. We assume that all nodes are either routers or networks (i.e., there are no "bridges," "hubs," etc.).*

Call your program `routed` *(as in* routing daemon*). Once started in command line, it awaits instructions and performs actions:*

(1) `add rt` ⟨routers⟩
This command adds routers to the routing table, where ⟨routers⟩ *is a comma separated list of (positive) integers and integer ranges. That is,* ⟨routers⟩ *can be* `6,9,10-13,4,8` *which would include routers*

    `rt4,rt6,rt8,rt9,rt10,rt11,rt12,rt13`

Your program should be robust enough to accept any such legal sequence (including a single router), and to return an error message if the command attempts to add a router that already exists (but other valid routers in the list ⟨routers⟩ *should be added regardless).*

(2) `del rt` ⟨routers⟩
Deletes routers given in ⟨routers⟩*. If the command attempts to delete a router that does not exist, an error message should be returned; we want robustness: routers that exist should be deleted, while attempting to delete non-existent routers should return an error message (specifying the "offending" routers). The program should not stop after displaying an error message.*

(3) `add nt` ⟨networks⟩
Add networks as specified in ⟨networks⟩*; same format as for adding routers. So for example "*`add nt 89`*" would result in the addition of* `nt89`*. The handling of errors should be done analogously to the case of adding routers.*

(4) `del nt` ⟨networks⟩
Deletes networks given in ⟨networks⟩*.*

(5) `con` $x\ y\ z$
Connect node x and node y, where x, y are existing routers and networks (for example, $x =$ `rt8` *and $y =$* `rt90`*, or $x =$* `nt76` *and $y =$* `rt1`*) and z is the cost of the connection. If x or y does not exist an error message should be returned. Note that the network is directed; that is, the following two commands are not equivalent:*

"`con rt3 rt5 1`" *and* "`con rt5 rt3 1.`"
It is important to note that two networks cannot be connected directly; an attempt to do so should generate an error message. If a connection between x and y already exists, it is updated with the new cost z.

(6) `display`
This command displays the routing table, i.e., the link-state database. *For example, the result of adding* `rt3`, `rt5`, `nt8`, `nt9` *and giving the commands* "`con rt5 rt3 1`" *and* "`con rt3 nt8 6`" *would display the following routing table:*

```
     rt3  rt5  nt8  nt9
rt3       1
rt5
nt8  6
nt9
```

Note that (according to the `RFC 2338`, *describing OSPF Version 2) we read the table as follows: "column first, then row." Thus, the table says that there is a connection from* `rt5` *to* `rt3`, *with cost 1, and another connection from* `rt3` *to* `nt8`, *with cost 6.*

(7) `tree` x
This commands computes the tree of shortest paths, with x as the root, from the link-state database. Note that x must be a router in this case. The output should be given as follows:

$$
\begin{aligned}
w_1 &: x, v_1, v_2, \ldots, v_n, y_1 \\
&: \textit{no path to } y_2 \\
w_3 &: x, u_1, u_2, \ldots, u_m, y_3 \\
&\vdots
\end{aligned}
$$

where w_1 is the cost of the path (the sum of the costs of the edges), from x to y_1, with v_i's the intermediate nodes (i.e., the "hops") to get from x to y_1. Every node y_j in the database should be listed; if there is no path from x to y_j it should say so, as in the above example output.
Following the example link-state database in the explanation of the `display` *command, the output of executing the command* "`tree rt5`" *would be:*

```
1 : rt5,rt3
7 : rt5,rt3,nt8
  : no path to nt9
```

Just as it is done in the OSPF standard, the path-tree should be computed with Dijkstra's greedy algorithm.
Finally, there may be several paths of the same value between two nodes; in that case, explain in the comments in your program how does your scheme select one of them.

(8) `quit`
Kills the daemon.

2.3.4 *Huffman codes*

One more important instance of a greedy solution is given by the Huffman algorithm, which is a widely used and effective technique for loss-less data compression. Huffman's algorithm uses a table of the frequencies of occurrences of the characters to build an optimal way of representing each character as a binary string. See §16.3 in [Cormen *et al.* (2009)] for details, but the following example illustrates the key idea.

Suppose that we have a string s over the alphabet $\{\mathtt{a}, \mathtt{b}, \mathtt{c}, \mathtt{d}, \mathtt{e}, \mathtt{f}\}$, and $|s| = 100$. Suppose also that the characters in s occur with the frequencies $44, 14, 11, 17, 8, 6$, respectively. As there are six characters, if we were using fixed-length binary codewords to represent them we would require three bits, and so 300 characters to represent the string.

Instead of a fixed-length encoding we want to give frequent characters a short codeword and infrequent characters a long codeword. We consider only codes in which no codeword is also a prefix of some other codeword. Such codes are called *prefix codes*; there is no loss of generality in restricting attention to prefix codes, as it is possible to show that any code can always be replaced with a prefix code that is at least as good.

Encoding and decoding is simple with a prefix code; to encode we just concatenate the codewords representing each character of the file. Since no codeword is a prefix of any other, the codeword that begins an encoded string is unambiguous, and so decoding is easy.

A prefix code can be given with a binary tree where the leaves are labeled with a character and its frequency, and each internal node is labeled with the sum of the frequencies of the leaves in its subtree. See figure 2.6. We construct the code of a character by traversing the tree starting at the root, and writing a 0 for a left-child and a 1 for a right-child.

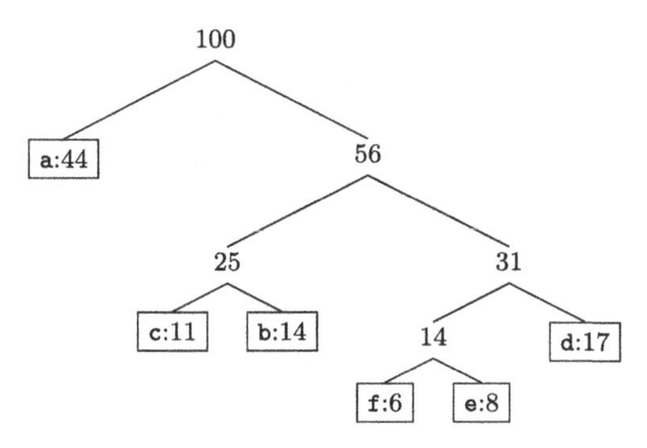

Fig. 2.6 Binary tree for the variable-length prefix code.

Let Σ be an alphabet of n characters and let $f : \Sigma \longrightarrow \mathbb{N}$ be the frequencies function. The Huffman algorithm builds a tree T corresponding to the optimal code in a bottom-up manner. It begins with a set of $|\Sigma|$ leaves and performs a sequence of $|\Sigma| - 1$ "merging" operations to create the final tree. At each step, the two least-frequent objects are merged together; the result of a merge of two objects is a new object whose frequency is the sum of the frequencies of the two objects that were merged.

Algorithm 2.5 Huffman

$n \leftarrow |\Sigma|$; $Q \leftarrow \Sigma$
for $i = 1..n-1$ **do**
  allocate a new node z
  left$[z] \leftarrow x =$ extract-min(Q)
  right$[z] \leftarrow y =$ extract-min(Q)
  $f(z) \leftarrow f(x) + f(y)$
  insert z in Q
end for

Problem 2.33. *Consider a file consisting of ASCII 100 characters, with the following frequencies:*

character	a	b	c	d	e	f	g	h
frequency	40	15	12	10	8	6	5	4

Using the standard ASCII encoding this file requires 800 bits. Compute a variable length prefix encoding for this file, and compute the total number of bits when using that encoding.

Problem 2.34. *Write a Python program that takes as input a text file, over, say, the ASCII alphabet, and uses Huffman's algorithm to compress it into a binary string. The compressed file should include a header containing the mapping of characters to bit strings, so that a properly compressed file can be decompressed. Your Python program should be able to do both: compress and decompress. Compare your solution to standard compression tools such as* `gzip`[2].

2.4 Answers to selected problems

Problem 2.2. A leaf is a vertex with one outgoing edge; suppose there is no leaf. Pick a vertex, take one of its outgoing edges. As each vertex has at least two adjacent edges, we keep going arriving at one edge, and leaving by the other. As there are finitely many edges we must eventually form a cycle. Contradiction.

We now show by induction on n that a tree with n nodes must have exactly $n-1$ edges. Basis case: $n = 1$, so the tree consists of a single node, and hence it has no edges; $n-1 = 1-1 = 0$ edges. Induction step: suppose that we have a tree with $n+1$ nodes. Pick a leaf and the edge that connects it to the rest of the tree. Removing this leaf and its edge results in a tree with n nodes, and hence—by induction hypothesis—with $n-1$ edges. Thus, the entire tree has $(n-1)+1 = n$ edges, as required.

Problem 2.4. We prove this by induction, with the basis case $n = 3$ (since a graph—without multiple edges between the same pair of nodes—cannot have a cycle with less than 3 nodes). If $n = 3$, and there are more than $n-1 = 2$ edges, there must be exactly 3 edges. So the graph is a cycle (a "triangle"). Induction step: consider a graph with $n+1$ many nodes ($n \geq 3$), and at least $n+1$ many edges. If the graph has a node with zero or one edges adjacent to it, then by removing that node (and its edge, if there is one), we obtain a graph with n nodes and at least n edges, and so—by induction hypothesis—the resulting graph has a cycle, and so the original graph also has a cycle. Otherwise, all nodes have at least two edges adjacent to it. Suppose v_0 is such a node, and $(v_0, x), (v_0, y)$ are two edges. Remove v_0 from the graph, and remove the edges $(v_0, x), (v_0, y)$ and replace them

[2] `gzip` implements the Lempel-Ziv-Welch (LZW) algorithm, which is a lossless data compression algorithm, available on UNIX platforms. It takes as input any file, and outputs a compressed version with the `.gz` extension. It is described in RFCs 1951 and 1952.

by the single edge (x, y). Again—by induction hypothesis—there must be a cycle in the resulting graph. But then there must be a cycle in the original graph as well. (Note that there are $n+1$ nodes, so after removing v_0 there are n nodes, and $n \geq 3$.)

Problem 2.11. Let e be any edge in $T_2 - T_1$. We must prove the existence of $e' \in T_1 - T_2$ such that $(T_1 \cup \{e\}) - \{e'\}$ is a spanning tree. Since $e \notin T_1$, by adding e to T_1 we obtain a cycle (by lemma 2.3, which is proved in problem 2.4). A cycle has at least 3 edges (the graph G has at least 3 nodes, since otherwise it could not have two distinct spanning trees!). So in this cycle, there is an edge e' not in T_2. The reason is that if every edge e' in the cycle did belong to T_2, then T_2 itself would have a cycle. By removing e', we break the cycle but the resulting graph, $(T_1 \cup \{e\}) - \{e'\}$, is still connected and of size $|T_1| = |T_2|$, i.e., the right size for a tree, so it must be acyclic (for otherwise, we could get rid of some edge, and have a spanning tree of size smaller than T_1 and T_2—but all spanning trees have the same size), and therefore $(T_1 \cup \{e\}) - \{e'\}$ is a spanning tree.

Problem 2.13. Let T be any MCST for a graph G. Reorder the edges of G by costs, just as in Kruskal's algorithm. For any block of edges of the same cost, put those edges which appear in T before all the other edges in that block. Now prove the following loop invariant: the set of edges S selected by the algorithm with the initial ordering as described is always a subset of T. Initially $S = \emptyset \subseteq T$. In the induction step, $S \subseteq T$, and S' is the result of adding one more edge to S. If $S' = S$ there is nothing to do, and if $S' = S \cup \{e\}$, then we need to show that $e \in T$. Suppose that it isn't. Let T' be the result of Kruskal's algorithm, which we know to be a MCST. By the exchange lemma, we know that there exists an $e' \notin T'$ such that $T \cup \{e\} - \{e'\}$ is a ST, and since T was a MCST, then $c(e') \leq c(e)$, and hence e' was considered *before* e. Since e' is not in T', it was rejected, so it must have created a cycle in S, and hence in T—contradiction. Thus $S \cup \{e\} \subseteq T$.

Problem 2.19. Here is the trace of the algorithm; note that we modify the optimal solution only as far as it is necessary to preserve the extension property.

$S^1 =$ | 1 | 0 | 0 | 0 | $\qquad S^1_{\text{opt}} =$ | 1 | 4 | 5 | 8 |

$S^2 =$ | 1 | 3 | 0 | 0 | $\qquad S^2_{\text{opt}} =$ | 1 | 3 | 5 | 8 |

$S^3 =$ | 1 | 3 | 0 | 5 | $\qquad S^3_{\text{opt}} =$ | 1 | 3 | 8 | 5 |

$S^4 =$ | 1 | 3 | 6 | 5 | $\qquad S^4_{\text{opt}} =$ | 1 | 3 | 6 | 5 |

Problem 2.21. Since $S' = S$ and $S'_{\text{opt}} = S_{\text{opt}}$ we must show that S is extendable into S_{opt} with jobs in $\{i+1, i+2, \ldots, n\}$. Since job i could not be scheduled in S, and S_{opt} extends S (i.e., S_{opt} has all the jobs that S had, and perhaps more), it follows that i could not be in S_{opt} either, and so i was not necessary in extending S into S_{opt}.

Problem 2.23. why 1. To show that S'_{opt} is feasible, we have to show that no job is scheduled twice, and no job is scheduled after its deadline. The first is easy, because S_{opt} was feasible. For the second we argue like this: the job that was at time t_0 is now moved to $t_1 < t_0$, so certainly if t_0 was before its deadline, so is t_1. The job that was at time t_1 (job i) has now been moved forward to time t_0, but we are working under the assumption that job i was scheduled (at this point) in slot t_0, so $t_0 \leq d_i$, and we are done. **why 2.** S'_{opt} extends S' because S_{opt} extended S, and the only difference is positions t_1 and t_0. They coincide in position t_0 (both have i), so we only have to examine position t_1. But $S(t_1) = 0$ since $S_{\text{opt}}(t_1) = i$, and S does not schedule job i at all. Since the only difference between S and S' is in position t_0, it follows that $S'(t_1) = 0$, so it does not matter what $S'_{\text{opt}}(t_1)$ is, it will extend S'. **why 3.** They schedule the same set of jobs, so they must have the same profit. **why 4.** Suppose $t_1 > t_0$. Since S_{opt} extends S, it follows that $S(t_1) = 0$. Since $S_{\text{opt}}(t_1) = i$, it follows that $t_1 \leq d_i$. But then, the algorithm would have scheduled i in t_1, not in t_0.

The fact that $j \neq 0$ is used in the last sentence of the proof of claim 2.22, where we conclude a contradiction from $S(t_2) = j \neq S_{\text{opt}}(t_2)$. If j were 0 then it could very well be that $S(t_2) = j = 0$ but $S_{\text{opt}}(t_2) \neq 0$.

Problem 2.27. Define a *promising list* to be one that can be extended to an optimal list of coins. Now show that L is promising is a loop invariant. Basis case: Initially, L is empty, so any optimal solution extends L. Hence L is promising. Induction step: Assume that L is promising, and show that L continues being promising after one more execution of the loop: Suppose L is promising, and $s < N$. Let L' be the list that extends L to the optimal solution, i.e. $L, L' = L_{\text{opt}}$. Let x be the largest item in C such that $s + x \leq N$. **Case (a)** $x \in L'$. Then $L' = x, L''$, so that L, x can be extended to the optimal solution L_{opt} by L''. **Case (b)** $x \notin L'$. We show that this case is not possible. To this end we prove the following claim:

Claim: If $x \notin L'$, then there is a sub-list L_0 of L' such that $x =$ sum of elements in L_0.

Proof of claim: Let B be the smallest number such that $B \geq x$, and some sub-list of L' sums to B. Let this sub-list be $\{e_1, e_2, \ldots, e_l\}$, where $e_i \leq e_{i+1}$ (i.e. the elements are in non-decreasing order). Since x is the largest coin that fits in $N - s$, and the sum of the coins in L' is $N - s$, it follows that every coin in L' is $\leq x$. Since $e_l \neq x$ (as $x \notin L'$), it follows that $l > 1$. Let $D = x - (e_2 + \ldots + e_l)$. By definition of B we know that $D > 0$. Each of the numbers $x, e_2, \ldots, e_l$ is divisible by e_1 (to see this note that all the coins are powers of p, i.e. in the set $\{1, p, p^2, \ldots, p^n\}$, and $e_l < x$ so $e_1 < x$). Thus $D \geq e_1$. On the other hand $x \leq e_1 + e_2 + \ldots + e_l$, so we also know that $D \leq e_1$, so in fact $D = e_1$. Therefore $x = e_1 + e_2 + \ldots + e_l$, and we are done. (end proof of claim)

Thus $\{e_1, e_2, \ldots, e_l\}$ can be replaced by the single coin x. If $l = 1$, then $x = e_1 \in L'$, which is a contradiction. If $l > 1$, then

$$L, x, L' - \{e_1, e_2, \ldots, e_l\}$$

sums up to N, but it has less coins than $L, L' = L_{\text{opt}}$ which is a contradiction. Thus case (b) is not possible.

Problem 2.28. See algorithm 2.6

Algorithm 2.6 Solution to problem 2.28.

$w(e_1) \geq w(e_2) \geq \ldots \geq w(e_m)$
$M \longleftarrow \emptyset$
for $i : 1..m$ **do**
  if $M \cup \{e_i\}$ does *not* contain two edges with a common vertex **then**
    $M \longleftarrow M \cup \{e_i\}$
  end if
end for

Problem 2.30. Let M_{opt} be an optimal matching. Define "M is promising" to mean that M can be extended to M_{opt} with edges that have not been considered yet. We show that "M is promising" is a loop invariant of our algorithm. The result will follow from this (it will also follows that there is a *unique* max matching). Basis case: $M = \emptyset$, so it is certainly promising. Induction step: Assume M is promising, and let M' be M after considering edge e_i. We show that: $e_i \in M' \iff e_i \in M_{\text{opt}}$.

[$\Longrightarrow$] Assume that $e_i \in M'$, since the weights are distinct, and powers of 2, $w(e_i) > \sum_{j=i+1}^{m} w(e_j)$ (to see why this holds, see problem 1.1), so unless $e_i \in M_{\text{opt}}$, $w(M_{\text{opt}}) < w$ where w is the result of algorithm.

[$\Longleftarrow$] Assume that $e_i \in M_{\text{opt}}$, so $M \cup \{e_i\}$ has no conflict, so the algorithm would add it.

Problem 2.31. This problem refers to Dijkstra's algorithm for the shortest path; for more background see §24.3, page 658, in [Cormen *et al.* (2009)] and §4.4, page 137, in [Kleinberg and Tardos (2006)]. The proof is simple: define S to be promising if for all the nodes v in S, $d(v)$ is indeed the shortest distance from s to v. We now need to show by induction on the number of iterations of the algorithm that "S is promising" is a loop invariant. The basis case is $S = \{s\}$ and $d(s) = 0$, so it obviously holds. For the induction step, suppose that v is the node just added, so $S' = S \cup \{v\}$. Suppose that there is a shorter path in G from s to v; call this path p (so p is just a sequence of nodes, starting at s and finishing at v). Since p starts inside S (at s) and finishes outside S (at v), it follows that there is an edge (a, b) such that a, b are consecutive nodes on p, where a is in S and b is in $V - S$. Let $c(p)$ be the cost of path p, and let $d'(v)$ be the value the algorithm found; we have $c(p) < d'(v)$. We now consider two cases: $b = v$ and $b \neq v$, and see that both yield a contradiction. Thus, no such path p exists. If $b = v$, then the algorithm would have used a instead of u. If $b \neq v$, then the cost of the path from s to b is even smaller than $c(p)$, so the algorithm would have added b instead of v.

2.5 Notes

Any book on algorithms has a chapter on greedy algorithms. For example, chapter 16 in [Cormen *et al.* (2009)] or chapter 4 in [Kleinberg and Tardos (2006)].

We also point out that there is a profound connection between a mathematical structure called *matroids* and greedy algorithms. A matroid, also known as an *independence structure*, captures the notion of "independence," just like the notion of independence in linear algebra.

A matroid M is a pair (E, I), where E is a finite set and I is a collection of subsets of E (called the *independent sets*) with the following three properties:

(i) The empty set is in I, i.e., $\emptyset \in I$.
(ii) Every subset of an independent set is also independent, i.e., if $x \subseteq y$, then $y \in I \Rightarrow x \in I$.
(iii) If x and y are two independent sets, and x has more elements than y, then there exists an element in x which is not in y that when added to y still gives an independent set. This is called the the *independent set exchange property*.

The last property is of course reminiscent of our Exchange lemma, lemma 2.10.

A good way to understand the meaning of this definition is to think of E as a set of vectors (in $\mathbb{R}^n$) and I all the subsets of E consisting of linearly independent vectors; check that all three properties hold.

For a review of the connection between matroids and greedy algorithms see [Papadimitriou and Steiglitz (1998)], chapter 12, "Spanning Trees and Matroids."

For a study of which optimization problems can be optimally or approximately solved by "greedy-like" algorithms see [Allan Borodin (2003)].

A well known algorithm for computing a maximum matching in a bipartite graph is the Hopcroft-Karp algorithm; see, for example, [Cormen *et al.* (2009)]. This algorithm runs in polynomial time (i.e., efficiently), but it is not greedy—the greedy approach seems to fail as section 2.3.2 insinuates.

Chapter 3

Divide and Conquer

Divide et impera—divide and conquer—was a Roman military strategy that consisted in securing command by breaking a large concentration of power into portions that alone were weaker, and methodically dispatching those portions one by one. This is the idea behind divide and conquer algorithms: take a large problem, divide it into smaller parts, solve those parts *recursively*, and combine the solutions to those parts into a solution to the whole.

The paradigmatic example of a divide and conquer algorithm is merge sort, where we have a large list of items to be sorted; we break it up into two smaller lists (divide), sort those recursively (conquer), and then combine those two sorted lists into one large sorted list. We present this algorithm in section 3.1. We also present a divide and conquer algorithm for binary integer multiplication—section 3.2, and graph reachability—section 3.3.

The divide and conquer approach is often used in situations where there is a brute force/exhaustive search algorithm that solves the problem, but the divide and conquer algorithm improves the running time. This is, for example, the case of binary integer multiplication. The last example in this chapter is a divide and conquer algorithm for reachability (Savitch's algorithm) that minimizes the use of memory, rather than the running time.

In order to analyze the use of resources (whether time or space) of a recursive procedure we must solve recurrences; see, for example, [Rosen (2007)] or [Cormen *et al.* (2009)] for the necessary background—"the master method" for solving recurrences. We provide a short discussion in the Notes section at the end of this chapter.

Recall that we discussed analyzing recursive algorithms in section 1.3.5.

3.1 Mergesort

Suppose that we have two lists of numbers that are already sorted. That is, we have a list $a_1 \le a_2 \le \cdots \le a_n$ and $b_1 \le b_2 \le \cdots \le b_m$. We want to combine those two lists into one long sorted list $c_1 \le c_2 \le \cdots \le c_{n+m}$. Algorithm 3.1 does the job.

Algorithm 3.1 Merge two lists

Pre-condition: $a_1 \le a_2 \le \cdots \le a_n$ and $b_1 \le b_2 \le \cdots \le b_m$

$p_1 \longleftarrow 1$; $p_2 \longleftarrow 1$; $i \longleftarrow 1$
while $i \le n + m$ **do**
    if $a_{p_1} \le b_{p_2}$ **then**
        $c_i \longleftarrow a_{p_1}$
        $p_1 \longleftarrow p_1 + 1$
    else
        $c_i \longleftarrow b_{p_1}$
        $p_2 \longleftarrow p_2 + 1$
    end if
    $i \longleftarrow i + 1$
end while

Post-condition: $c_1 \le c_2 \le \cdots \le c_{n+m}$

The mergesort algorithm sorts a given list of numbers by first dividing them into two lists of length $\lceil n/2 \rceil$ and $\lfloor n/2 \rfloor$, respectively, then sorting each list recursively, and finally combining the results using algorithm 3.1.

Algorithm 3.2 Mergesort

Pre-condition: A list of integers $a_1, a_2, \ldots, a_n$

1: $L \longleftarrow a_1, a_2, \ldots, a_n$
2: **if** $|L| \le 1$ **then**
3:     **return** L
4: **else**
5:     $L_1 \longleftarrow$ first $\lceil n/2 \rceil$ elements of L
6:     $L_2 \longleftarrow$ last $\lfloor n/2 \rfloor$ elements of L
7:     **return** Merge(Mergesort(L_1), Mergesort(L_2))
8: **end if**

Post-condition: $a_{i_1} \le a_{i_2} \le \cdots \le a_{i_n}$

In algorithm 3.2, line 1 sets L to be the list of the input numbers $a_1, a_2, \ldots, a_n$. These are integers, not necessarily ordered. Line 2 checks if L is not empty or consists of a single element; if that is the case, then the list is already sorted—this is where the recursion "bottoms out," by returning the same list. Otherwise, in line 5 we let L_1 consist of the first $\lceil n/2 \rceil$ elements of L and L_2 consist of the last $\lfloor n/2 \rfloor$ elements of L.

Problem 3.1. *Show that* $L = L_1 \cup L_2$.

In section 1.3.5 of the Preliminaries chapter we have shown how to use the theory of fixed points to prove the correctness of recursive algorithms. For us this will remain a theoretical tool, as it is not easy to come up with the least fixed point that interprets a recursion. We are going to give natural proofs of correctness using induction.

Problem 3.2. *Prove the correctness of the Mergesort algorithm.*

Let $T(n)$ bound the running time of the mergesort algorithm on lists of length n. Clearly,

$$T(n) \leq T(\lceil n/2 \rceil) + T(\lfloor n/2 \rfloor) + cn,$$

where cn, for some constant c, is the cost of the merging of the two lists (algorithm 3.1). Furthermore, the asymptotic bounds are not affected by the floors and the ceils, and so we can simply say that $T(n) \leq 2T(n/2) + cn$. Thus, $T(n)$ is bounded by $O(n \log n)$.

Problem 3.3. *Implement mergesort in Python. Assume that the input is given as a list of numbers, given in a textfile.*

3.2 Multiplying numbers in binary

Consider the example of multiplication of two binary numbers, using the junior school algorithm, given in figure 3.1.

This school multiplication algorithm is very simple. To multiply x times y, where x, y are two numbers in binary, we go through y from right to left; when we encounter a 0 we write a row of as many zeros as $|x|$, the length of x. When we encounter a 1 we copy x. When we move to the next bit of y we shift by one space to the left. At the end we produce the familiar "stairs" shape—see s_1, s_2, s_3, s_4 in figure 3.1 (henceforth, figure 3.1 is our running example of binary multiplication).

	1	2	3	4	5	6	7	8
x					1	1	1	0
y					1	1	0	1
s_1					1	1	1	0
s_2				0	0	0	0	
s_3			1	1	1	0		
s_4		1	1	1	0			
$x \times y$	1	0	1	1	0	1	1	0

Fig. 3.1 Multiply 1110 times 1101, i.e., 14 times 13.

Once we obtain the "stairs," we go back to the top step (line s_1) and to its right-most bit (column 8). To obtain the product we add all the entries in each column with the usual carry operation. For example, column 5 contains two ones, so we write a 0 in the last row (row $x \times y$) and carry over 1 to column 4. It is not hard to see that multiplying two n-bit integers takes $O(n^2)$ primitive bit operations.

We now present a divide and conquer algorithm that takes only $O(n^{\log 3}) \approx O(n^{1.59})$ operations. The speed-up obtained from the divide and conquer procedure appears slight—but the improvement does become substantial as n grows very large.

Let x and y be two n-bit integers. We break them up into two smaller $n/2$-bit integers as follows:

$$x = (x_1 \cdot 2^{n/2} + x_0),$$
$$y = (y_1 \cdot 2^{n/2} + y_0).$$

Thus x_1 and y_1 correspond to the high-order bits of x and y, respectively, and x_0 and y_0 to the low-order bits of x and y, respectively. The product of x and y appears as follows in terms of those parts:

$$\begin{aligned} xy &= (x_1 \cdot 2^{n/2} + x_0)(y_1 \cdot 2^{n/2} + y_0) \\ &= x_1y_1 \cdot 2^n + (x_1y_0 + x_0y_1) \cdot 2^{n/2} + x_0y_0. \end{aligned} \tag{3.1}$$

A divide and conquer procedure appears surreptitiously. To compute the product of x and y we compute the four products $x_1y_1, x_1y_0, x_0y_1, x_0y_0$, *recursively*, and then we combine them as in (3.1) to obtain xy.

Let $T(n)$ be the number of operations that are required to compute the product of two n-bit integers using the divide and conquer procedure that arises from (3.1). Then

$$T(n) \leq 4T(n/2) + cn, \tag{3.2}$$

since we have to compute the four products $x_1y_1, x_1y_0, x_0y_1, x_0y_0$ (this is where the $4T(n/2)$ factor comes from), and then we have to perform three additions of n-bit integers (that is where the factor cn, where c is some constant, comes from). Notice that we do not take into account the product by 2^n and $2^{n/2}$ (in (3.1)) as they simply consist in shifting the binary string by an appropriate number of bits to the left (n for 2^n and $n/2$ for $2^{n/2}$). These shift operations are inexpensive, and can be ignored in the complexity analysis.

When we solve the standard recurrence given by (3.2), we can see that $T(n) \leq O(n^{\log 4}) = O(n^2)$, so it seems that we have gained nothing over the brute force procedure.

It appears from (3.1) that we have to make four recursive calls; that is, we need to compute the four multiplications $x_1y_1, x_1y_0, x_0y_1, x_0y_0$. But we can get away with only three multiplications, and hence four recursive calls: x_1y_1, x_0y_0 and $(x_1 + x_0)(y_1 + y_0)$; the reason being that

$$(x_1y_0 + x_0y_1) = (x_1 + x_0)(y_1 + y_0) - (x_1y_1 + x_0y_0). \tag{3.3}$$

See figure 3.2 for a comparison of the cost of operations.

	multiplications	additions	shifts
Method (3.1)	4	3	2
Method (3.3)	3	4	2

Fig. 3.2 Reducing the number of multiplications by one increase the number of additions by one—something has to give. But, as multiplications are more expensive, the trade is worth it.

Algorithm 3.3 implements the idea given by (3.3).

Algorithm 3.3 Recursive binary multiplication

Pre-condition: Two n-bit integers x and y

1: $x \longleftarrow (x_1 \cdot 2^{\lceil n/2 \rceil} + x_0)$
2: $y \longleftarrow (y_1 \cdot 2^{\lceil n/2 \rceil} + y_0)$
3: $z_1 \longleftarrow$ Multiply$(x_1 + x_0, y_1 + y_0)$
4: $z_2 \longleftarrow$ Multiply(x_1, y_1)
5: $z_3 \longleftarrow$ Multiply(x_0, y_0)
6: **return** $z_2 \cdot 2^n + (z_1 - z_2 - z_3) \cdot 2^{\lceil n/2 \rceil} + z_3$

Note that in lines 1 and 2 of the algorithm, we break up x and y into two parts x_1, x_0 and y_1, y_0, respectively, where x_1, y_1 consist of the $\lfloor n/2 \rfloor$

high order bits, and x_0, y_0 consist of the $\lceil n/2 \rceil$ low order bits.

Problem 3.4. *Prove the correctness of algorithm 3.3.*

Algorithm 3.3 clearly takes $T(n) \leq 3T(n/2) + dn$ operations. Thus, the running time is $O(n^{\log 3}) \approx O(n^{1.59})$—to see this read the discussion on solving recurrences in the Notes section of this chapter.

Problem 3.5. *Implement the binary multiplication algorithm in Python. Assume that the input is given in the command line as two strings of zeros and ones.*

3.3 Savitch's algorithm

In this section we are going to give a divide and conquer solution to the graph reachability problem. Recall the graph-theoretic definitions that were given at the beginning of section 2.1. Here we assume that we have a (directed) graph G, and we want to establish whether there is a path from some node s to some node t; note that we are not even searching for a shortest path (as in section 2.3.3 or in section 4.2); we just want to know if node t is reachable from node s.

As a twist on minimizing the running time of algorithms, we are going to present a very clever divide and conquer solution that reduces drastically the amount of space, i.e., memory. Savitch's algorithm solves directed reachability in space $O(\log^2 n)$, where n is the number of vertices in the graph. This is remarkable, as $O(\log^2 n)$ bits of memory is very little space indeed, for a graph with n vertices! We assume that the graph is presented as an $n \times n$ adjacency matrix (see page 39), and so it takes exactly n^2 bits of memory—that is, "work memory," which we use to implement the stack.

It might seem futile to commend an algorithm that takes $O(\log^2 n)$ bits of space when the input itself requires n^2 bits. If the input already takes so much space, what benefit is there to requiring small space for the computations? The point is that the input does not have to be presented in its entirety. The graph may be given *implicitly*, rather than *explicitly*. For example, the "graph" $G = (V, E)$ may be the entire World Wide Web, where V is the set of all web pages (at a given moment in time) and there is an edge from page x to page y if there is hyperlink in x pointing to y. We may be interested in the existence of a path in the WWW, and we can query the pages and their links piecemeal without maintaining the representation of the entire WWW in memory. The sheer size of the WWW is such that it may be beneficial to know that we only require as much space as the square of the logarithm of the number of web pages.

Incidentally, we are not saying that Savitch's algorithm is the ideal solution to the "WWW hyperlink connectivity problem"; we are simply giving an example of an enormous graph, and an algorithm that uses very little working space with respect to the size of the input.

Define the Boolean predicate $\mathrm{R}(G, u, v, i)$ to be true iff there is a path in G from u to v of length at most 2^i. The key idea is that if a path exists from u to v, then any such path must have a mid-point w; a seemingly trivial observation that nevertheless inspires a very clever recursive procedure. In other words there exist paths of distance at most 2^{i-1} from u to w and from w to v, i.e.,

$$\mathrm{R}(G, u, v, i) \iff (\exists w)[\mathrm{R}(G, u, w, i-1) \wedge \mathrm{R}(G, w, v, i-1)]. \tag{3.4}$$

Algorithm 3.4 computes the predicate $\mathrm{R}(G, u, v, i)$ based on the recurrence given in (3.4). Note that in algorithm 3.4 we are computing $\mathrm{R}(G, u, v, i)$; the recursive calls come in line 9 where we compute $\mathrm{R}(G, u, w, i-1)$ and $\mathrm{R}(G, w, v, i-1)$.

Algorithm 3.4 Savitch

```
1: if i = 0 then
2:         if u = v then
3:                 return T
4:         else if (u, v) is an edge then
5:                 return T
6:         end if
7: else
8:         for every vertex w do
9:                 if R(G, u, w, i − 1) and R(G, w, v, i − 1) then
10:                        return T
11:                end if
12:        end for
13: end if
14: return F
```

Problem 3.6. *Show that algorithm 3.4 is correct (i.e., it computes* $\mathrm{R}(G, u, v, i)$ *correctly) and it requires at most* $i \cdot s$ *space, where* s *is the* *number of bits required to keep record of a single node. Conclude that it requires $O(\log^2 n)$ space on a graph G with n nodes.*

Problem 3.7. *Algorithm 3.4 truly uses very little space to establish connectivity in a graph. But what is the* time *complexity of this algorithm?*

Problem 3.8. *Implement Savitch's algorithm in Python.*

3.4 Further examples and exercises

3.4.1 *Extended Euclid's algorithm*

We revisit an old friend from section 1.3, namely the extended Euclid's algorithm—see problem 1.21, and the corresponding solution on page 29 containing algorithm 1.15. We present a recursive version, as algorithm 3.5, where the algorithm returns three values, and hence we use the notation $(x, y, z) \longleftarrow (x', y', z')$ as a convenient shorthand for $x \longleftarrow x'$, $y \longleftarrow y'$ and $z \longleftarrow z'$. Note the interesting similarity between algorithm 1.15 and the Gaussian lattice reduction—algorithm 1.7.

Algorithm 3.5 Extended Euclid's algorithm (recursive)

Pre-condition: $m > 0, n \geq 0$

1: $a \longleftarrow m$; $b \longleftarrow n$
2: **if** $b = 0$ **then**
3:     **return** $(a, 1, 0)$
4: **else**
5:     $(d, x, y) \longleftarrow \text{Euclid}(b, \text{rem}(a, b))$
6:     **return** $(d, y, x - \text{div}(a, b) \cdot y)$
7: **end if**

Post-condition: $mx + ny = d = \gcd(m, n)$

Problem 3.9. *Show that algorithm 3.5 works correctly.*

Problem 3.10. *Implement the extended Euclid's algorithm (algorithm 3.5) in Python. Assume that the input is given in the command line, as two integers.*

3.4.2 *Finite automata*

In this section we are going to present a little bit of material from automata theory. It is not our aim to present this vast field—the interested reader is directed to [Sipser (2006)] or [Kozen (2006)]—but we want to introduce some notation that will be used in this section, as well as in section 4.6.3.

An *alphabet* is a finite, non-empty set of distinct symbols, denoted usually by Σ. For example, $\Sigma = \{0, 1\}$, the usual binary alphabet, or $\Sigma = \{a, b, c, \ldots, z\}$, the usual lower-case letters of the English alphabet. A *string*, also called *word*, is a finite ordered sequence of symbols chosen from some alphabet. For example, 010011101011 is a string over the binary alphabet. The notation $|w|$ denotes the *length* of the string w, e.g., $|010011101011| = 12$. The *empty string*, ε, $|\varepsilon| = 0$, is in any Σ by default. Σ^k is the set of strings over Σ of length exactly k, for example, if $\Sigma = \{0, 1\}$, then

$$\begin{aligned}
\Sigma^0 &= \{\varepsilon\} \\
\Sigma^1 &= \Sigma \\
\Sigma^2 &= \{00, 01, 10, 11\}
\end{aligned}$$

The set Σ^* is called *Kleene's star* of Σ, and it is the set of all strings over Σ. Note that $\Sigma^* = \Sigma^0 \cup \Sigma^1 \cup \Sigma^2 \cup \ldots$, while $\Sigma^+ = \Sigma^1 \cup \Sigma^2 \cup \ldots$. If x, y are strings, and $x = a_1 a_2 \ldots a_m$ and $y = b_1 b_2 \ldots b_n$ then their *concatenation* is just their juxtaposition, i.e., $x \cdot y = a_1 a_2 \ldots a_m b_1 b_2 \ldots b_n$. We often write xy, instead of $x \cdot y$. Note $w\varepsilon = \varepsilon w = w$. A *language* L is a collection of strings over some alphabet Σ, i.e., $L \subseteq \Sigma^*$. For example,

$$L = \{\varepsilon, 01, 0011, 000111, \ldots\} = \{0^n 1^n | n \geq 0\} \tag{3.5}$$

Note that $\{\varepsilon\} \neq \emptyset$; one is the language consisting of the single string ε, and the other is the empty language.

A *Deterministic Finite Automaton (DFA)* a tuple $A = (Q, \Sigma, \delta, q_0, F)$ where: Q is a finite set of *states*; Σ is a finite alphabet, which is a set of input symbols; $\delta : Q \times \Sigma \longrightarrow Q$ is a transition function—this is "the program" that runs the DFA, so to speak. Given $q \in Q, a \in \Sigma$, we obtain $\delta(q, a) = p \in Q$. There is also a start state, sometimes called an *initial state*, q_0, and a set of final (accepting) states, F.

To see whether A accepts a string w, we "run" A on $w = a_1 a_2 \ldots a_n$ as follows: $\delta(q_0, a_1) = q_1$, $\delta(q_1, a_2) = q_2$, until $\delta(q_{n-1}, a_n) = q_n$. We say that A *accepts* w iff $q_n \in F$, i.e., if q_n is one of the final (accepting) states. More precisely: A accepts w if there exists a sequence of states $r_0, r_1, \ldots, r_n$,

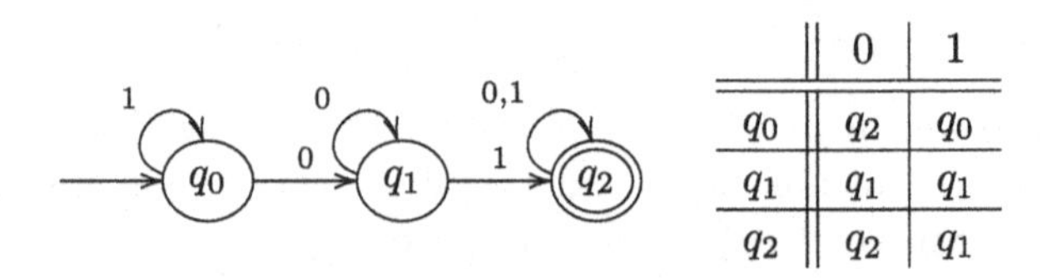

	0	1
q_0	q_2	q_0
q_1	q_1	q_1
q_2	q_2	q_1

Fig. 3.3 Transition diagram and transition table

where $n = |w|$, such that $r_0 = q_0$, $\delta(r_i, w_{i+1}) = r_{i+1}$ where $i = 0, 1, \dots, n-1$ and w_j is the j-th symbol of w, and $r_n \in F$.

As an example, consider $L = \{w |\ w \text{ is of the form } x01y \in \Sigma^* \}$ where $\Sigma = \{0, 1\}$. We want to specify a DFA $A = (Q, \Sigma, \delta, q_0, F)$ that accepts all and only the strings in L. We let $\Sigma = \{0, 1\}$, $Q = \{q_0, q_1, q_2\}$, and $F = \{q_1\}$; see figure 3.3 for the transition function, where it is specified both as a transition diagram and a transition table.

Extended Transition Function (ETF) given δ, its ETF is $\hat{\delta}$ defined inductively: basis case: $\hat{\delta}(q, \varepsilon) = q$; induction step: if $w = xa$, $w, x \in \Sigma^*$ and $a \in \Sigma$, then

$$\hat{\delta}(q, w) = \hat{\delta}(q, xa) = \delta(\hat{\delta}(q, x), a).$$

Thus $\hat{\delta} : Q \times \Sigma^* \longrightarrow Q$, and $w \in L(A) \iff \hat{\delta}(q_0, w) \in F$. Here $L(A)$ is the set of all those strings (and only those) which are accepted by A. The language of a DFA A is defined as $L(A) = \{w | \hat{\delta}(q_0, w) \in F\}$ We say that a language L is *regular* if there exists a DFA A such that $L = L(A)$.

We are often interested in finding a minimal DFA for a given language. We say that two states are *equivalent* if for all strings w, $\hat{\delta}(p, w)$ is accepting $\iff$ $\hat{\delta}(q, w)$ is accepting. If two states are not equivalent, they are *distinguishable.*

We have a recursive (divide-and-conquer) procedure for finding pairs of distinguishable states. First, if p is accepting and q is not, then $\{p, q\}$ is a pair of distinguishable states. This is the "bottom" case of the recursion. If $r = \delta(p, a)$ and $s = \delta(q, a)$, where $a \in \Sigma$ and $\{r, s\}$ are already found to be distinguishable, then $\{p, q\}$ are distinguishable; this is the recursive case. We want to formalize this with the so called *table filling algorithm*, which is a recursive algorithm for finding distinguishable pairs of states.

Problem 3.11. *Design the recursive table filling algorithm. Prove that in your algorithm, if two states are not distinguished by the algorithm, then the two states are equivalent.*

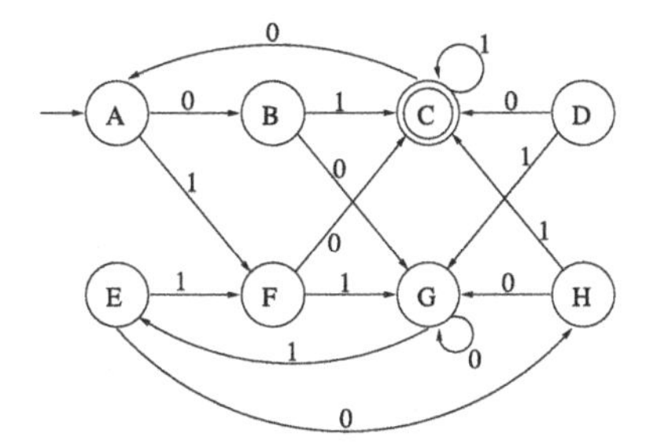

	A	B	C	D	E	F	G
B	×						
C	×	×					
D	×	×	×				
E		×	×	×			
F	×	×	×		×		
G	×	×	×	×	×	×	
H	×		×	×	×	×	×

Fig. 3.4 An example of a DFA and the corresponding table. Distinguishable states are marked by "×"; the table is only filled below the diagonal, as it is symmetric.

We now use the table filling algorithm to show both the equivalence of automata and to minimize them. Suppose D_1, D_2 are two DFAs. To see if they are equivalent, i.e., $L(D_1) = L(D_2)$, run the table-filling algorithm on their "union", and check if $q_0^{D_1}$ and $q_0^{D_2}$ are equivalent.

Note that the equivalence of states is an equivalence relation (see appendix B). We can use this fact to minimize DFAs. For a given DFA, we run the Table Filling Algorithm, to find all the equivalent states, and hence all the equivalence classes. We call each equivalence class a *block*. In the example in figure 3.4, the blocks would be:

$$\{E, A\}, \{H, B\}, \{C\}, \{F, D\}, \{G\}$$

The states within each block are equivalent, and the blocks are disjoint.

We now build a minimal DFA with states given by the blocks as follows: $\gamma(S, a) = T$, where $\delta(p, a) \in T$ for $p \in S$. We must show that γ is well defined; suppose we choose a different $q \in S$. Is it still true that $\delta(q, a) \in T$? Suppose not, i.e., $\delta(q, a) \in T'$, so $\delta(p, a) = t \in T$, and $\delta(q, a) = t' \in T'$. Since $T \neq T'$, $\{t, t'\}$ is a distinguishable pair. But then so is $\{p, q\}$, which contradicts that they are both in S.

Problem 3.12. *Show that we obtain a minimal DFA from this procedure.*

Problem 3.13. *Implement the minimization algorithm in Python. Assume that the input is given as a transition table, where the alphabet is fixed to be* $\{0, 1\}$*, and the rows represent states, where the first row stands for the initial state. Indicate the rows that correspond to accepting states with a special symbol, for example,* $*$.

Note that with this convention you do not need to label the rows and columns of the input, except for the $*$ *denoting the accepting states. Thus, the transition table given in figure 3.4 would be represented as follows:*

```
  2 0
  1 1
* 2 1
```

3.5 Answers to selected problems

Problem 3.1. It is enough to show that $\lceil n/2 \rceil + \lfloor n/2 \rfloor = n$. If n is even, then $\lceil n/2 \rceil = \lfloor n/2 \rfloor = n/2$, and $n/2+n/2 = n$. If n is odd, then $n = 2k+1$, and so $\lceil n/2 \rceil = k+1$ while $\lfloor n/2 \rfloor = k$ and $(k+1)+k = 2k+1 = n$.
Problem 3.2. We must show that given a list of integers $L = a_1, a_2, \ldots, a_n$, the algorithm returns a list L', which consists of the numbers in L in non-decreasing order. The recursion itself suggest the right induction; we use the CIP (see page 3). If $|L| = 1$ then $L' = L$ and we are done. Otherwise, $|L| = n > 1$, and we obtain two lists L_1 and L_2 (of lengths $\lceil n/2 \rceil$ and $n - \lceil n/2 \rceil$), which, by induction hypothesis, are returned ordered. Now it remains to prove the correctness of the merging procedure, algorithm 3.1, which can also be done by induction.
Problem 3.7. $O(2^{\log^2 n}) = O(n^{\log n})$, so the time complexity of Savitch's algorithm is super-polynomial, and so not very good.
Problem 3.9. First note that the second argument decreases at each recursive call, but by definition of remainder, it is non-negative. Thus, by the LNP, the algorithm terminates. We prove partial correctness by induction on the value of the second argument. In the basis case $n = 0$, so in line 1 $b \longleftarrow n = 0$, so in line 2 $b = 0$ and the algorithm terminates in line 3 and returns $(a, 1, 0) = (m, 1, 0)$, so $mx + ny = m \cdot 1 + n \cdot 0 = m$ while $d = m$, and so we are done.

In the induction step we assume that the recursive procedure returns correct values for all pairs of arguments where the second argument is $< n$ (thus, we are doing complete induction). We have that

$$\begin{aligned}(d, x, y) \longleftarrow & \text{ Extended-Euclid}(b, \text{rem}(a, b)) \\ = & \text{ Extended-Euclid}(n, \text{rem}(m, n)),\end{aligned}$$

from lines 1 and 5. Note that $0 \leq \text{rem}(m, n) < n$, and so we can apply the induction hypothesis and we have that:

$$n \cdot x + \text{rem}(m, n) \cdot y = d = \gcd(n, \text{rem}(m, n)).$$

First note that by problem 1.18 we have that $d = \gcd(m, n)$. Now we work on the left-hand side of the equation. We have:

$$\begin{aligned}
&n \cdot x + \text{rem}(m, n) \cdot y \\
=&n \cdot x + (m - \text{div}(m, n) \cdot n) \cdot y \\
=&m \cdot y + n \cdot (x - \text{div}(m, n) \cdot y) \\
=&m \cdot y + n \cdot (x - \text{div}(a, b) \cdot y)
\end{aligned}$$

and we are done as this is what is returned in line 6.

Problem 3.11. Here we present a generic proof for the "natural algorithm" that you should have designed for filling out the table. We use an argument by contradiction with the Least Number Principle (LPN). Let $\{p, q\}$ be a distinguishable pair, for which the algorithm left the corresponding square empty, and furthermore, of all such "bad" pairs $\{p, q\}$ has a shortest distinguishing string w. Let $w = a_1 a_2 \ldots a_n$, $\hat{\delta}(p, w)$ is accepting while $\hat{\delta}(q, w)$ is not. First, $w \neq \varepsilon$, as then p, q would have been found to be distinguishable in the basis case of the algorithm. Let $r = \delta(p, a_1)$ and $s = \delta(q, a_1)$. Then, $\{r, s\}$ are distinguished by $w' = a_2 a_3 \ldots a_n$, and since $|w'| < |w|$, they were found out by the algorithm. But then $\{p, q\}$ would have been found in the next stage.

Problem 3.12. Consider a DFA A on which we run the procedure to obtain M. Suppose that there exists an N such that $L(N) = L(M) = L(A)$, and N has fewer states than M. Run the Table Filling Algorithm on M, N together (renaming the states, so they do not have states in common). Since $L(M) = L(N)$ their initial states are indistinguishable. Thus, each state in M is indistinguishable from at least one state in N. But then, two states of M are indistinguishable from the same state of $N \ldots$

3.6 Notes

For a full discussion of Mergesort and binary multiplication, see §5.1 and §5.5, respectively, in [Kleinberg and Tardos (2006)]. For more background on Savitch's algorithm (section 3.3) see theorem 7.5 in [Papadimitriou (1994)], §8.1 in [Sipser (2006)] or theorem 2.7 in [Kozen (2006)].

The reachability problem is ubiquitous in computer science. Suppose that we have a graph G with n nodes. In section 2.3.3 we presented a $O(n^2)$ time greedy algorithm for reachability, due to Dijkstra. In this chapter, in section 3.3, we presented a divide and conquer algorithm that requires $O(\log^2 n)$ space, due to Savitch . In section 4.2 we will present a dynamic programming algorithm that computes the shortest paths for all the pairs of nodes in the graph—it is due to Floyd and takes time $O(n^3)$. In subsection 4.2.1 we present another dynamic algorithm due to Bellman and Ford (which can cope with edges of negative weight). In 2005, Reingold showed that undirected reachability can be computed in space $O(\log n)$; see [Reingold (2005)] for this remarkable, but difficult, result. Note that Reingold's algorithm works for undirected graphs only.

See chapter 7 in [Rosen (2007)] for an introduction to solving recurrence relations, and §4.5, pages 93–103, in [Cormen *et al.* (2009)] for a very thorough discussion of the "master method" for solving recurrences. Here we include a very short discussion; we want to solve recurrences of the following form:

$$T(n) = aT(n/b) + f(n), \tag{3.6}$$

where $a \geq 1$ and $b > 1$ are constants and $f(n)$ is an asymptotically positive function—meaning that there exists an n_0 such that $f(n) > 0$ for all $n \geq n_0$. There are three cases for solving such a recurrence.

Case 1 is $f(n) = O(n^{\log_b a - \varepsilon})$ for some constant $\varepsilon > 0$; in this case we have that $T(n) = \Theta(n^{\log_b a})$. Case 2 is $f(n) = \Theta(n^{\log_b a} \log^k n)$ with $k \geq 0$; in this case we have that $T(n) = \Theta(n^{\log_b a} \log^{k+1} n)$. Finally, Case 3 is $f(n) = \Omega(n^{\log_b a + \varepsilon})$ with $\varepsilon > 0$, and $f(n)$ satisfies the *regularity condition*, namely $af(n/b) \leq cf(n)$ for some constant $c < 1$ and all sufficiently large n; in this case $T(n) = \Theta(f(n))$.

For example, the recurrence that appears in the analysis of mergesort is $T(n) = 2T(n/2) + cn$, so $a = 2$ and $b = 2$, and so $\log_b a = \log_2 2 = 1$, and so we can say that $f(n) = \Theta(n^{\log_b a} \log^k n) = \Theta(n \log n)$, i.e., $k = 1$ in Case 2, and so $T(n) = \Theta(n \log n)$ as was pointed out in the analysis.

Chapter 4

Dynamic Programming

Dynamic programming is an algorithmic technique that is closely related to the divide and conquer approach we saw in the previous chapter. However, while the divide and conquer approach is essentially recursive, and so "top down," dynamic programming works "bottom up."

A dynamic programming algorithm creates an array of related but simpler subproblems, and then, it computes the solution to the big complicated problem by using the solutions to the easier subproblems which are stored in the array. We usually want to maximize profit or minimize cost.

There are three steps in finding a dynamic programming solution to a problem: (i) Define a class of subproblems, (ii) give a recurrence based on solving each subproblem in terms of simpler subproblems, and (iii) give an algorithm for computing the recurrence.

4.1 Longest monotone subsequence problem

Input: $d, a_1, a_2, \ldots, a_d \in \mathbb{N}$.
Output: $L =$ length of the longest monotone non-decreasing subsequence.

Note that a subsequence need not be consecutive, that is $a_{i_1}, a_{i_2}, \ldots, a_{i_k}$ is a monotone subsequence provided that

$$1 \leq i_1 < i_2 < \ldots < i_k \leq d,$$
$$a_{i_1} \leq a_{i_2} \leq \ldots \leq a_{i_k}.$$

For example, the length of the longest monotone subsequence (henceforth LMS) of $\{4, 6, 5, 9, 1\}$ is 3.

We first define an array of subproblems: $R(j) =$ length of the longest monotone subsequence which ends in a_j. The answer can be extracted from array R by computing $L = \max_{1 \leq j \leq n} R(j)$.

The next step is to find a recurrence. Let $R(1) = 1$, and for $j > 1$,

$$R(j) = \begin{cases} 1 & \text{if } a_i > a_j \text{ for all } 1 \leq i < j \\ 1 + \max_{1 \leq i < j}\{R(i) | a_i \leq a_j\} & \text{otherwise} \end{cases}.$$

We finish by writing an algorithm that computes R; see algorithm 4.1.

Algorithm 4.1 Longest monotone subsequence (LMS)

```
R(1) ← 1
for j : 2..d do
    max ← 0
    for i : 1..j − 1 do
        if R(i) > max and a_i ≤ a_j then
            max ← R(i)
        end if
    end for
    R(j) ← max +1
end for
```

Problem 4.1. *Once we have computed all the values of the array R, how could we build an actual monotone non-decreasing subsequence of length L?*

Problem 4.2. *What would be the appropriate pre/post-conditions of the above algorithms? Prove correctness with an appropriate loop invariant.*

Problem 4.3. *Consider the following variant of the Longest Monotone Subsequence problem. The input is still $d, a_1, a_2, \ldots, a_d \in \mathbb{N}$, but the output is the the length of the longest subsequence of $a_1, a_2, \ldots, a_d$, where any two consecutive members of the subsequence differ by at most 1. For example, the longest such subsequence of $\{7, 6, 1, 4, 7, 8, 20\}$ is $\{7, 6, 7, 8\}$, so in this case the answer would be 4. Give a dynamic programming solution to this problem.*

Problem 4.4. *Implement algorithm 4.1 in Python; your algorithm should take an extra step parameter, call it s, where, just as in problem 4.3, any two consecutive members of the subsequence differ by at most s, that is, $|a_{i_j} - a_{i_{j+1}}| \leq s$, for any $1 \leq j < k$.*

4.2 All pairs shortest path problem

Input: Directed graph $G = (V, E)$, $V = \{1, 2, \ldots, n\}$, and a cost function $C(i, j) \in \mathbb{N}^+ \cup \{\infty\}$, $1 \leq i, j \leq n$, $C(i, j) = \infty$ if (i, j) is not an edge.
Output: An array D, where $D(i, j)$ the length of the shortest directed path from i to j.

Recall that we have defined undirected graphs in section 2.1; a *directed graph* (or *digraph*) is a graph where the edges have a direction, i.e., the edges are arrows. Also recall that in section 2.3.3 we have given a greedy algorithm for computing the shortest paths from a designated node s to all the nodes in an (undirected) graph.

Problem 4.5. *Construct a family of graphs $\{G_n\}$, where G_n has $O(n)$ many nodes, and exponentially many paths, that is $\Omega(2^n)$ paths. Conclude, therefore, that an exhaustive search is not a feasible solutions to the "all pairs shortest path problem."*

Define an array of subproblems: let $A(k, i, j)$ be the length of the shortest path from i to j such that all *intermediate* nodes on the path are in $\{1, 2, \ldots, k\}$. Then $A(n, i, j) = D(i, j)$ will be the solution. The convention is that if $k = 0$ then $\{1, 2, \ldots, k\} = \emptyset$.

Define a recurrence: we first initialize the array for $k = 0$ as follows: $A(0, i, j) = C(i, j)$. Now we want to compute $A(k, i, j)$ for $k > 0$. To design the recurrence, notice that the shortest path between i and j either includes k or does not. Assume we know $A(k-1, r, s)$ for all r, s. Suppose node k is not included. Then, obviously, $A(k, i, j) = A(k-1, i, j)$. If, on the other hand, node k occurs on a shortest path, then it occurs exactly once, so $A(k, i, j) = A(k-1, i, k) + A(k-1, k, j)$. Therefore, the shortest path length is obtained by taking the minimum of these two cases:

$$A(k, i, j) = \min\{A(k-1, i, j), A(k-1, i, k) + A(k-1, k, j)\}.$$

Write an algorithm: it turns out that we only need space for a two-dimensional array $B(i, j) = A(k, i, j)$, because to compute $A(k, *, *)$ from $A(k-1, *, *)$ we can overwrite $A(k-1, *, *)$.

Our solution is algorithm 4.2, known as Floyd's algorithm (or the Floyd-Warshall algorithm). It is remarkable as it runs in time $O(n^3)$, where n is the number of vertices, while there may be up to $O(n^2)$ edges in such a graph. In lines 1–5 we initialize the array B, i.e., we set it equal to C. Note that before line 6 is executed, it is the case that $B(i, j) = A(k-1, i, j)$ for all i, j.

Algorithm 4.2 Floyd

```
1: for i : 1..n do
2:     for j : 1..n do
3:         B(i, j) ⟵ C(i, j)
4:     end for
5: end for
6: for k : 1..n do
7:     for i : 1..n do
8:         for j : 1..n do
9:             B(i, j) ⟵ min{B(i, j), B(i, k) + B(k, j)}
10:        end for
11:    end for
12: end for
13: return D ⟵ B
```

Problem 4.6. *Why does the overwriting method in algorithm 4.2 work? The worry is that $B(i, k)$ or $B(k, j)$ may have already been updated (if $k < j$ or $k < i$). However, the overwriting* does *work; explain why. We could have avoided a 3-dimensional array by keeping two 2-dimensional arrays instead, and then overwriting would not be an issue at all; how would that work?*

Problem 4.7. *In algorithm 4.2, what are appropriate pre and post-conditions? What is an appropriate loop invariant?*

Problem 4.8. *Implement Floyd's algorithm in Python, using the two dimensional array and the overwriting method.*

4.2.1 *Bellman-Ford algorithm*

Suppose that we want to find the shortest path from s to t, in a directed graph $G = (V, E)$, where edges have non-negative costs. Let $\text{OPT}(i, v)$ denote the minimal cost of an i-path from v to t, where an i-path is a path that uses at most i edges. Let p be an optimal i-path with cost $\text{OPT}(i, v)$; if no such p exists we adopt the convention that $\text{OPT}(i, v) = \infty$.

If p uses $i-1$ edges, then $\text{OPT}(i, v) = \text{OPT}(i-1, v)$, and if p uses i edges, and the first edge is $(v, w) \in E$, then $\text{OPT}(i, v) = c(v, w) + \text{OPT}(i-1, w)$, where $c(v, w)$ is the cost of edge (v, w). This gives us the recursive formula, for $i > 0$: $\text{OPT}(i, v) = \min\{\text{OPT}(i-1, v), \min_{w \in V}\{c(v, w) + \text{OPT}(i-1, w)\}\}$.

Problem 4.9. *Implement Bellman-Ford's algorithm in Python.*

4.3 Simple knapsack problem

Input: $w_1, w_2, \ldots, w_d, C \in \mathbb{N}$, where C is the knapsack's capacity.
Output: $\max_S\{K(S)|K(S) \leq C\}$, where $S \subseteq [d]$ and $K(S) = \sum_{i \in S} w_i$.

This is an NP-hard[1] problem, which means that we cannot expect to find a polynomial time algorithm that works in general. We give a dynamic programming solution that works for relatively small C; note that for our method to work the inputs $w_1, \ldots, w_d, C$ must be (non-negative) integers. We often abbreviate the name "simple knapsack problem" with SKS.

Define an array of subproblems: we consider the first i weights (i.e., $[i]$) summing up to an *intermediate* weight limit j. We define a Boolean array R as follows:

$$R(i,j) = \begin{cases} \mathsf{T} & \text{if } \exists S \subseteq [i] \text{ such that } K(S) = j \\ \mathsf{F} & \text{otherwise} \end{cases},$$

for $0 \leq i \leq d$ and $0 \leq j \leq C$. Once we have computed all the values of R we can obtain the solution M as follows: $M = \max_{j \leq C}\{j | R(d,j) = \mathsf{T}\}$.

Define a recurrence: we initialize $R(0,j) = \mathsf{F}$ for $j = 1, 2, \ldots, C$, and $R(i,0) = \mathsf{T}$ for $i = 0, 1, \ldots, d$.

We now define the recurrence for computing R, for $i, j > 0$, in a way that hinges on whether we include object i in the knapsack. Suppose that we do *not* include object i. Then, obviously, $R(i,j) = \mathsf{T}$ iff $R(i-1,j) = \mathsf{T}$. Suppose, on the other hand, that object i *is* included. Then it must be the case that $R(i,j) = \mathsf{T}$ iff $R(i-1, j-w_i) = \mathsf{T}$ and $j - w_i \geq 0$, i.e., there is a subset $S \subseteq [i-1]$ such that $K(S)$ is exactly $j - w_i$ (in which case $j \geq w_i$). Putting it all together we obtain the following recurrence for $i, j > 0$:

$$R(i,j) = \mathsf{T} \iff R(i-1,j) = \mathsf{T} \vee (j \geq w_i \wedge R(i-1, j-w_i) = \mathsf{T}). \quad (4.1)$$

Figure 4.1 on the next page summarizes the computation of the recurrence.

We finally design algorithm 4.3 that uses the same space saving trick as algorithm 4.2; it employs a one-dimensional array $S(j)$ for keeping track of a two-dimensional array $R(i,j)$. This is done by overwriting $R(i,j)$ with $R(i+1,j)$.

In algorithm 4.3, in line 1 we initialize the array for $i = j = 0$. In lines 2–4 we initialize the array for $i = 0$ and $j \in \{1, 2, \ldots, C\}$. Note that

[1] NP is the class of problems solvable in polynomial time on a *non-deterministic*Turing machine. A problem P is NP-hard if every problem in NP is reducible to P in polynomial time, that is, every problem in NP can be efficiently restate in terms of P. When a problem is NP-hard this is an indication that it is probably *intractable*, i.e., it cannot be solved efficiently in general. For more information on this see any book on complexity, for example [Papadimitriou (1994); Sipser (2006); Soltys (2009)].

R	0	$\cdots$	$j-w_i$	$\cdots$	j	$\cdots$	C
0	T	F···F	F	F···F	F	F···F	F
	T ⋮ T						
$i-1$	T		**c**		**b**		
i	T				**a**		
	T ⋮ T						
d	T						

Fig. 4.1 The recurrence given by the equivalence (4.1) can be interpreted as follows: we place a T in the square labeled with **a** if and only if at least one of the following two conditions is satisfied: there is a T in the position right above it, i.e., in the square labeled with **b** (if we can construct j with the first $i-1$ weights, surely we can construct j with he first i weights), or there is a T in the square labeled with **c** (if we can construct $j - w_i$ with the first $i - 1$ weights, surely we can construct j with the first i weights). Also note that to fill the square labeled with **a** we only need to look at two squares, and neither of those two squares is to the right; this will be important in the design of the algorithm (algorithm 4.3).

after each execution of the i-loop (line 5) it is the case that $S(j) = R(i, j)$ for all j.

Problem 4.10. *We are using a one dimensional array to keep track of a two dimensional array, but the overwriting is not a problem; explain why.*

Problem 4.11. *The assertion $S(j) = R(i, j)$ can be proved by induction on the number of times the i-loop in algorithm 4.3 is executed. This assertion implies that upon termination of the algorithm, $S(j) = R(d, j)$ for all j. Prove this formally, by giving pre/post-conditions, a loop invariant, and a standard proof of correctness.*

Problem 4.12. *Construct an input for which algorithm 4.3 would make an error if the inner loop "for decreasing $j : C..1$" (line 6) were changed to "for $j : 1..C$."*

Problem 4.13. *Implement algorithm 4.3 in Python.*

Algorithm 4.3 is a nice illustration of the powerful idea of *program refinement.* We start with the idea of computing $R(i, j)$ for all i, j. We then

Algorithm 4.3 Simple knapsack (SKS)

```
1: S(0) ⟵ T
2: for j : 1..C do
3:         S(j) ⟵ F
4: end for
5: for i : 1..d do
6:         for decreasing j : C..1 do
7:                 if (j ≥ w_i and S(j − w_i) = T) then
8:                         S(j) ⟵ T
9:                 end if
10:        end for
11: end for
```

realize that we only really need two rows in memory; to compute row i we only need to look up row $i-1$. We then take it further and see that by updating row i from right to left we do not require row $i-1$ at all—we can do it *mise en place*. By starting with a robust idea, and by successively trimming the unnecessary fat, we obtain a slick solution.

But how slick is our dynamic programming solution in terms of the complexity of the problem? That is, how many steps does it take to compute the solution *proportionally* to the size of the input? We must construct a $d \times C$ table and fill it in, so the time complexity of our solution is $O(d \cdot C)$. This seems acceptable at first glance, but we were saying in the introduction to this section that SKS is an NP-hard problem; what gives?

The point is that the input is assumed to be given in binary, and to encode C in binary we require only $\log C$ bits, and so the number of columns (C) is in fact exponential in the size of the input $(C = 2^{\log C})$. On the other hand, d is the number of weights, and since those weights must be listed somehow, the size of the list of weights is certainly bigger than d (i.e., this list cannot be encoded—in general—with $\log d$ bits; it requires at least d bits).

All we can say is that if C is of size $O(d^k)$, for some constant k, then our dynamic programming solution works in polynomial time in the size of the input. In other words, we have an efficient solution for "small" values of C. Another way of saying this is that as long as $|C|$ (the size of the binary encoding of C) is $O(\log d)$ our solution works in polynomial time.

Problem 4.14. *Show how to construct the actual optimal set of weights once R has been computed.*

Problem 4.15. *Define a "natural" greedy algorithm for solving SKS; let* $\overline{M}$ *be the output of this algorithm, and let* M *be the output of the dynamic programming solution given in this section. Show that either* $\overline{M} = M$ *or* $\overline{M} > \frac{1}{2}C$.

Problem 4.15 introduces surreptitiously the concept of *approximation algorithms.* As was mentioned at the beginning of this section (see footnote on page 81), SKS is an example of an NP-hard problem, a problem for which we suspect there may be no efficient solution in the general case. That is, the majority of experts believe that any algorithm—attempting to solve SKS in the general case—on infinitely many inputs will take an inordinate number of steps (i.e., time) to produce a solution.

One possible compromise is to design an efficient algorithm that does not give an *optimal* solution—which may not even be required—but only a solution with some guarantees as to its closeness to an optimal solution. Thus, we merely *approximate* the optimal solution but at least our algorithm runs quickly. The study of such compromises is undertaken by the field of approximation algorithms.

Finally, in the section below we give a greedy solution to SKS in the particular case where the weights have a certain "increasing property." This is an example of a *promise* problem, where we can expect some convenient condition on the inputs; a condition that we need not check for, but assume that we have. Note that we have been using the term "promising" to prove the correctness of greedy algorithms—this is a different notion from that of a "promise problem."

4.3.1 *Dispersed knapsack problem*

Input: $w_1, \ldots, w_d, C \in \mathbb{N}$, such that $w_i \geq \sum_{j=i+1}^{d} w_j$ for $i = 1, \ldots, d-1$.
Output: $S_{\max} \subseteq [d]$ where $K(S_{\max}) = \max_{S \subseteq [d]} \{K(S) | K(S) \leq C\}$.

Problem 4.16. *Give a "natural" greedy algorithm which solves Dispersed Knapsack by filling in the blanks in algorithm 4.4.*

Problem 4.17. *Give a definition of what it means for an intermediate solution* S *in algorithm 4.4 to be "promising." Show that the loop invariant "*S *is promising" implies that the greedy algorithm gives the optimal solution. Finally, show that "*S *is promising" is a loop invariant.*

Algorithm 4.4 Dispersed knapsack

```
S ⟵ ∅
for i : 1..d do
    if ____________________ then
        ____________________
    end if
end for
```

4.3.2 *General knapsack problem*

Input: $w_1, w_2, \ldots, w_d, v_1, \ldots, v_d, C \in \mathbb{N}$
Output: $\max_{S \subseteq [d]}\{V(S)|K(S) \leq C\}$, $K(S) = \sum_{i \in S} w_i$, $V(S) = \sum_{i \in S} v_i$.

Thus, the general knapsack problem (which we abbreviate as GKS) has a positive integer value v_i besides each weight w_i, and the goal is to have as valuable a knapsack as possible, without exceeding C, i.e., the weight capacity of the knapsack.

More precisely, $V(S) = \sum_{i \in S} v_i$ is the total value of the set S of weights. The goal is to maximize $V(S)$, subject to the constraint that $K(S)$, which is the sum of the weights in S, is at most C. Note that SKS is a special case of GKS where $v_i = w_i$, for all $1 \leq i \leq d$.

To solve GKS, we start by computing the same Boolean array $R(i, j)$ that was used to solve SKS. Thus $R(i, j)$ ignores the values v_i, and only depends on the weights w_i. Next we define another array $V(i, j)$ that depends on the values v_i as follows:

$$V(i, j) = \max\{V(S)|S \subseteq [i] \text{ and } K(S) = j\}, \tag{4.2}$$

for $0 \leq i \leq d$ and $0 \leq j \leq C$.

Problem 4.18. *Give a recurrence for computing the array $V(i, j)$, using the Boolean array $R(i, j)$—assume that the array $R(i, j)$ has already been computed. Also, give an algorithm for computing $V(i, j)$.*

Problem 4.19. *If the definition of $V(i, j)$ given in (4.2) is changed so that we only require $K(S) \leq j$ instead of $K(S) = j$, then the Boolean array $R(i, j)$ is not needed in the recurrence. Give the recurrence in this case.*

4.4 Activity selection problem

Input: A list of activities $(s_1, f_1, p_1), \ldots, (s_n, f_n, p_n)$, where $p_i > 0$, $s_i < f_i$ and s_i, f_i, p_i are non-negative real numbers.
Output: A set $S \subseteq [n]$ of selected activities such that no two selected activities overlap, and the profit $P(S) = \sum_{i \in S} p_i$ is as large as possible.

An *activity* i has a fixed start time s_i, finish time f_i and profit p_i. Given a set of activities, we want to select a subset of non-overlapping activities with maximum total profit. A typical example of the activity selection problem is a set of lectures with fixed start and finish times that need to be scheduled in a single class room.

Define an array of subproblems: sort the activities by their finish times, $f_1 \leq f_2 \leq \ldots \leq f_n$. As it is possible that activities finish at the same time, we select the *distinct* finish times, and denote them $u_1 < u_2 < \ldots < u_k$, where, clearly, $k \leq n$. For instance, if we have activities finishing at times 1.24, 4, 3.77, 1.24, 5 and 3.77, then we partition them into four groups: activities finishing at times $u_1 = 1.24$, $u_2 = 3.77$, $u_3 = 4$, $u_4 = 5$.

Let u_0 be $\min_{1 \leq i \leq n} s_i$, i.e., the earliest start time. Thus,

$$u_0 < u_1 < u_2 < \ldots < u_k,$$

as it is understood that $s_i < f_i$. Define an array $A(0..k)$ as follows:

$$A(j) = \max_{S \subseteq [n]} \{P(S) | S \text{ is feasible and } f_i \leq u_j \text{ for each } i \in S\},$$

where S is *feasible* if no two activities in S overlap. Note that $A(k)$ is the maximum possible profit for all feasible schedules S.

Problem 4.20. *Give a formal definition of what it means for a schedule of activities to be feasible, i.e., express precisely that the activities in a set S "do not overlap."*

Define a recurrence for $A(0..k)$. In order to give such a recurrence we first define an auxiliary array $H(1..n)$ such that $H(i)$ is the index of the largest distinct finish time no greater than the start time of activity i. Formally, $H(i) = l$ if l is the largest number such that $u_l \leq s_i$. To compute $H(i)$, we need to search the list of distinct finish times. To do it efficiently, for each i, apply the binary search procedure that runs in logarithmic time in the length of the list of distinct finish times (try $l = \lfloor \frac{k}{2} \rfloor$ first). Since the length k of the list of distinct finish times is at most n, and we need to apply binary search for each element of the array $H(1..n)$, the time required to compute all entries of the array is $O(n \log n)$.

We initialize $A(0) = 0$, and we want to compute $A(j)$ given that we already have $A(0), \ldots, A(j-1)$. Consider $u_0 < u_1 < u_2 < \ldots < u_{j-1} < u_j$. Can we beat profit $A(j-1)$ by scheduling some activity that finishes at time u_j? Try all activities that finish at this time and compute maximum profit in each case. We obtain the following recurrence:

$$A(j) = \max\{A(j-1), \max_{1 \le i \le n} \{p_i + A(H(i)) \mid f_i = u_j\}\}, \tag{4.3}$$

where $H(i)$ is the greatest l such that $u_l \le s_i$. Consider the example given in figure 4.2.

Consider the example in figure 4.3. To see how the bottom row of the right-hand table in figure 4.3 was computed, note that according to the recurrence (4.3), we have:

$$A(2) = \max\{20, 30 + A(0), 20 + A(1)\} = 40,$$
$$A(3) = \max\{40, 30 + A(0)\} = 40.$$

Therefore, the maximum profit is $A(3) = 40$.

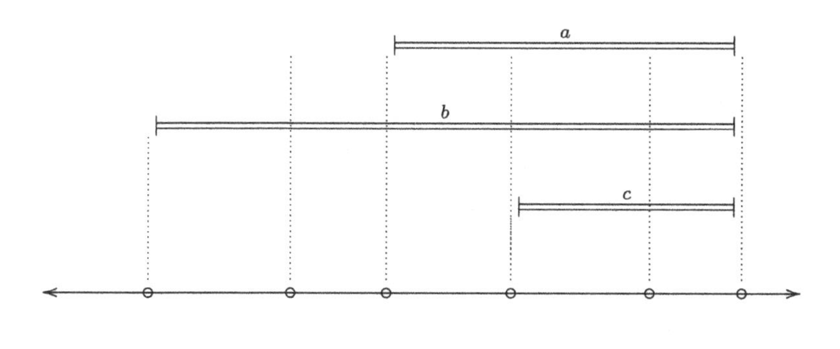

$s_b = u_{H(b)}$ $u_{H(a)}$ s_b $s_c = u_{H(c)}$ u_{j-1} u_j

Fig. 4.2 In this example we want to compute $A(j)$. Suppose that some activity finishing at time u_j must be scheduled in order to obtain the maximum possible profit. In this figure there are three activities that end at time u_j: a, b, c, given by $(s_a, f_a, p_a), (s_b, f_b, p_b), (s_c, f_c, p_c)$, respectively, where of course the assumption is that $u_j = f_a = f_b = f_c$. The question is which of these three activities must be selected. In order to establish this, we must look at each activity a, b, c in turn, and see what is the most profitable schedule that we can get if we *insist* that the given activity is scheduled. For example, if we insist that activity a be scheduled, we must see what is the most profitable schedule we can get where all other activities must finish by s_a, which effectively means that all other activities must finish by $u_{H(a)}$. Note that in this example we have that $u_{H(a)} < s_a$, but $u_{H(b)} = s_b$ and $u_{H(c)} = s_c$. When all is said, we must find which of the three values $p_a + A(H(a)), p_b + A(H(b)), p_c + A(H(c))$ is maximal.

Activity i:	1	2	3	4
Start s_i:	0	2	3	2
Finish f_i:	3	6	6	10
Profit p_i:	20	30	20	30
$H(i)$:	0	0	1	0

j:	0	1	2	3
u_j:	0	3	6	10
$A(j)$:	0	20	40	40

Fig. 4.3 An example with four activities.

Problem 4.21. *Write the algorithm.*

Problem 4.22. *Given that A has been computed, how do you find a set of activities S such that $P(S) = A(k)$? Hint: If $A(k) = A(k-1)$, then we know that no selected activity finishes at time u_k, so we go on to consider $A(k-1)$. If $A(k) > A(k-1)$, then some selected activity finishes at time u_k. How do we find this activity?*

Problem 4.23. *Implement in Python the dynamic programming solution to the "activity selection with profits problem." Your algorithm should compute the value of the most profitable set of activities, as well as output an explicit list of those activities.*

4.5 Jobs with deadlines, durations and profits

Input: A list of jobs $(d_1, t_1, p_1), \ldots, (d_n, t_n, p_n)$.
Output: A feasible schedule $C(1..n)$ such that the profit of C, denoted $P(C)$, is the maximum possible among feasible schedules.

In section 2.2 we considered the job scheduling problems for the case where each job takes unit time, i.e., each duration $d_i = 1$. We now generalize this to the case in which each job i has an arbitrary duration d_i, deadline t_i and profit p_i. We assume that d_i and t_i are positive integers, but the profit p_i can be a positive real number. We say that the schedule $C(1..n)$ is *feasible* if the following two conditions hold (let $C(i) = -1$ denote that job i is *not* scheduled, and so $C(i) \geq 0$ indicates that it it *is* scheduled, and note that we do allow jobs to be scheduled at time 0):

(1) if $C(i) \geq 0$, then $C(i) + d_i \leq t_i$; and,
(2) if $i \neq j$ and $C(i), C(j) \geq 0$, then

(a) $C(i) + d_i \leq C(j)$; or,
(b) $C(j) + d_j \leq C(i)$.

The first condition is akin to saying that each scheduled job finishes by its deadline and the second condition is akin to saying that no two scheduled jobs overlap. The goal is to find a feasible schedule $C(1..n)$ for the n jobs for which the profit $P(C) = \sum_{C(i) \geq 0} p_i$, the sum of the profits of the scheduled jobs, is maximized.

A job differs from an activity in that a job can be scheduled any time as long as it finishes by its deadline; an activity has a fixed start time and finish time. Because of the flexibility in scheduling jobs, it is "harder" to find an optimal schedule for jobs than to select an optimal subset of activities.

Note that job scheduling is "at least as hard as SKS." In fact an SKS instance $w_1, \ldots, w_n, C$ can be viewed as a job scheduling problem in which each duration $d_i = w_i$, each deadline $t_i = C$, and each profit $p_i = w_i$. Then the maximum profit of any schedule is the same as the maximum weight that can be put into the knapsack. This seemingly innocent idea of "at least as hard as" is in fact a powerful tool widely used in the field of computational complexity to compare the relative difficulty of problems. By restating a general instance of job scheduling as an instance of SKS we provided a *reduction* of job scheduling to SKS, and shown thereby that if one were able to solve job scheduling efficiently, one would automatically have an efficient solution to SKS.

To give a dynamic programming solution to the job scheduling problem, we start by sorting the jobs according to deadlines. Thus, we assume that $t_1 \leq t_2 \leq \ldots \leq t_n$.

It turns out that to define a suitable array A for solving the problem, we must consider all possible integer times t, $0 \leq t \leq t_n$ as a deadline for the first i jobs. It is not enough to only consider the specified deadline t_i given in the problem input. Thus define the array $A(i, t)$ as follows:

$$A(i,t) = \max \left\{ P(C) : \begin{array}{l} C \text{ is a feasible schedule} \\ \text{only jobs in } [i] \text{ are scheduled} \\ \text{all scheduled jobs finish by time } t \end{array} \right\}.$$

We now want to design a recurrence for computing $A(i, t)$. In the usual style, consider the two cases that either job i occurs or does not occur in the optimal schedule (and note that job i will not occur in the optimal schedule if $d_i > \min\{t_i, t\}$). If job i does not occur, we already know the optimal profit.

If, on the other hand, job i does occur in an optimal schedule, then we may as well assume that it is the last job (among jobs $\{1, \ldots, i\}$) to be scheduled, because it has the latest deadline. Hence we assume that job i is scheduled as late as possible, so that it finishes either at time t, or at time t_i, whichever is smaller, i.e., it finishes at time $t_{\min} = \min\{t_i, t\}$.

Problem 4.24. *In light of the discussion in the above two paragraphs, find a recurrence for* $A(i, t)$*.*

Problem 4.25. *Implement your solution in Python.*

4.6 Further examples and problems

4.6.1 *Consecutive subsequence sum problem*

Input: Real numbers $r_1, \ldots, r_n$
Output: For each consecutive subsequence of the form $r_i, r_{i+1}, \ldots, r_j$ let

$$S_{ij} = r_i + r_{i+1} + \cdots + r_j$$

where $S_{ii} = r_i$. Find $M = \max_{1 \leq i \leq j \leq n} S_{ij}$.

For example, in figure 4.4 we have a sample consecutive subsequence sum problem. There, the solution is $M = S_{35} = 3 + (-1) + 2 = 4$.

This problem can be solved in time $O(n^2)$ by systematically computing all of the sums S_{ij} and finding the maximum (there are $\binom{n}{2}$ pairs $i, j \leq n$ such that $i < j$). However, there is a more efficient dynamic programming solution which runs in time $O(n)$.

Define the array $M(1..n)$ by:

$$M(j) = \max\{S_{1j}, S_{2j}, \ldots, S_{jj}\}.$$

See figure 4.4 for an example.

Problem 4.26. *Explain how to find the solution* M *from the array* $M(1..n)$*.*

Problem 4.27. *Complete the four lines indicated in algorithm 4.5 for computing the values of the array* $M(1..n)$*, given* $r_1, r_2, \ldots, r_n$*.*

j	1	2	3	4	5	6	7
r_j	1	−5	3	−1	2	−8	3
$M(j)$	1	−4	3	2	4	−4	3

Fig. 4.4 An example of computing $M(j)$.

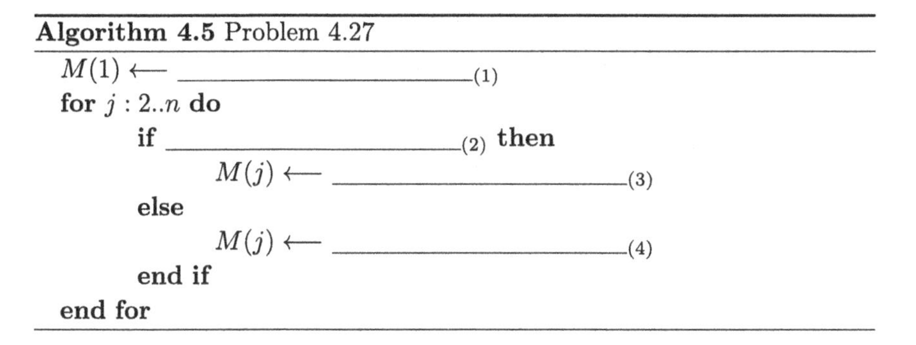

Algorithm 4.5 Problem 4.27

$M(1) \longleftarrow$ ______________________(1)
for $j : 2..n$ **do**
  if ______________________(2) **then**
    $M(j) \longleftarrow$ ______________________(3)
  else
    $M(j) \longleftarrow$ ______________________(4)
  end if
end for

4.6.2 *Regular expressions*

In section 3.4.2 we gave a very quick introduction to finite automata; our goal was to present the recursive table filling algorithm. As was mentioned earlier, the field of finite automata is a rich and vast area of study, and here we only mention it briefly. On page 72 we defined regular languages as those that can be described with a finite automaton.

We now present a different way of formalizing regular languages, using regular expressions, which are familiar to most people from searching text documents with utilities such as `grep` and `awk`. A *regular expression* is defined by structural induction as follows. Basis case: $a \in \Sigma, \varepsilon, \emptyset$. Induction step: If E, F are regular expressions, the so are $E + F, EF, (E)^*, (E)$.

Problem 4.28. *In the same vein that we defined the language $L(A)$ of a finite automaton A (page 72), we can define the language of a regular expression, following its definition by structural induction. What are $L(a), L(\varepsilon), L(\emptyset), L(E+F), L(EF), L(E^*)$?*

As an example, we give a regular expression for the set of strings of 0s and 1s *not* containing 101 as a substring:

$$(\varepsilon + 0)(1^* + 00^*0)^*(\varepsilon + 0)$$

We need the $(\varepsilon + 0)$ at the beginning and end to account for strings that start and end in zero (like $010, 01, 10$, etc.). By the way, the 1^* can be exchanged for just 1, since the star above the main bracket can generate any number of 1s at will.

Theorem 4.29. *A language is regular if it is given by some regular expression. Thus, finite automata and regular expressions capture the same class of languages.*

The proof of this theorem is not difficult, but outside the scope of this book. Here we only show one direction: we show how to convert an automaton into a regular expressions, since this procedure can be accomplished with an elegant dynamic programming algorithm.

Suppose A has n states. Let $R_{ij}^{(k)}$ denote the regular expression whose language is the set of strings w such that: w takes A from state i to state j with all intermediate states $\leq k$. The question is, what is R such that $L(R) = L(A)$? It conveniently turns out that $R = R_{1j_1}^{(n)} + R_{1j_2}^{(n)} + \cdots + R_{1j_k}^{(n)}$ where $F = \{j_1, j_2, \ldots, j_k\}$

We build $R_{ij}^{(k)}$ by induction on k. Basis case: $k = 0$,

$$R_{ij}^{(0)} = x + a_1 + a_2 + \cdots + a_k,$$

where $i \xrightarrow{a_l} j$ and $x = \emptyset$ if $i \neq j$ and $x = \varepsilon$ if $i = j$

Induction step: $k > 0$

$$R_{ij}^{(k)} = \underbrace{R_{ij}^{(k-1)}}_{\text{path does not visit } k} + \underbrace{R_{ik}^{(k-1)} \left(R_{kk}^{(k-1)}\right)^* R_{kj}^{(k-1)}}_{\text{visits } k \text{ at least once}}$$

As an example we convert a DFA that accepts only those strings that have 00 as a substring. The DFA is given by

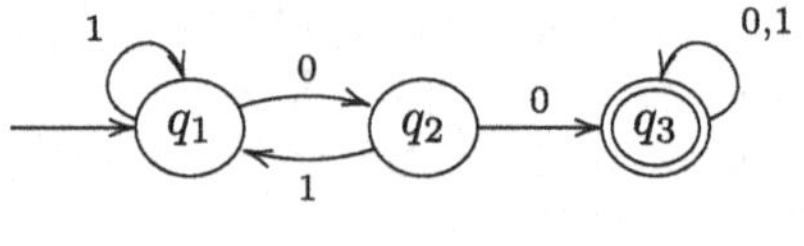

1 0 0,1
q_1 q_2 0 q_3
1

And $R_{11}^{(0)} = \varepsilon + 1$, $R_{12}^{(0)} = R_{23}^{(0)} = 0$, $R_{13}^{(0)} = R_{31}^{(0)} = R_{32}^{(0)} = \emptyset$ and $R_{21}^{(0)} = 1$ and $R_{33}^{(0)} = \varepsilon + 0 + 1$.

Problem 4.30. *Compute $R^{(1)}, R^{(2)}, R^{(3)}$, and give $R_{13}^{(3)}$.*

Problem 4.31. *Implement this translation in Python. To make this more challenging, you might consider implementing simplification rules; for example, you may simplify $(R^*)^*$ by R^*, and use the following identities: $\emptyset^* = \varepsilon$, $\varepsilon^* = \varepsilon$, $\emptyset + R = R + \emptyset = R$, $\emptyset R = R\emptyset = \emptyset$ and $\varepsilon R = R\varepsilon = R$, etc.*

4.6.3 *Context free grammars*

Recall from section 3.4.2 that an *alphabet* is a finite set of symbols, also known as *terminals* when the background is context free grammars, for example $\Sigma = \{0, 1\}$ or $\Sigma = \{a, b, c, \ldots, z\}$. Given an alphabet Σ, we denote the set of all finite strings over Σ as Σ^*. For example, if $\Sigma = \{0, 1\}$ then $\Sigma^* = \{\varepsilon, 0, 1, 00, 01, 10, 11, 000, 001, \ldots\}$, the set of all binary strings, where ε is the empty string. If w is a string, $|w|$ is the length of w, so $|\varepsilon| = 0$ and $|101001| = 6$. A *language* over Σ is a subset of Σ^*.

A *context free grammar* is a tuple $G = (V, T, R, S)$, where V is a (finite) set of *variables*, T is a (finite) set of terminals, R is a (finite) set of rules, and S denotes the *starting variable*. All rules are of the form $X \longrightarrow \alpha$ where $X \in V$ and $\alpha \in (V \cup T)^*$, i.e., α is a (finite) string of variables and terminals[2].

A *derivation* of a string w is denoted as: $S \Rightarrow \alpha_1 \Rightarrow \alpha_2 \Rightarrow \cdots \Rightarrow \alpha_k$, where $\alpha_k = w$ and each α_{i+1} is obtained from α_i by one of the rules, meaning that a variable X in α_i is replaced by α to obtain α_{i+1} if $X \longrightarrow \alpha$ was one of the rules of the grammar.

The set $L(G)$, called *the language of the grammar* G, is the set of all those strings in T^* which have G-derivations.

Sometimes, for the sake of succinctness, if there are several rules for the same variable X, for example $X \longrightarrow \gamma_1, \cdots, X \longrightarrow \gamma_k$, we simply write $X \longrightarrow \gamma_1 | \cdots | \gamma_k$ as if it were one rule, but the symbol "|" denotes that this is a family of rules for X.

For example, consider $G_{\text{pal}} = (\{S\}, \{0, 1\}, \{S \longrightarrow \varepsilon|0|1|0S0|1S1\}, S)$. In this grammar there is only one variable, S, which is also (necessarily) the starting variable, and there are five rules, which are represented in succinct form as a "single" rule $S \longrightarrow \varepsilon|0|1|0S0|1S1$, and there are two terminals $0, 1$.

The derivation

$$S \Rightarrow 0S0 \Rightarrow 01S10 \Rightarrow 01010,$$

shows that $01010 \in L(G_{\text{pal}})$. In fact, G_{pal} generates precisely the set of all palindromes[3] over 0,1, i.e., the set of all binary strings that read the same backwards as forwards.

[2]If the rules are of the form $\alpha \longrightarrow \beta$, where $|\alpha| \leq |\beta|$, then we have a *context sensitive grammar*; if there are no conditions on α and β then the resulting system is known as a *rewriting system*, and has the power of general algorithms—any algorithm in this book can be implemented as such a "rewriting system."

[3]We already encountered palindromes on page 10, and we have examined the correctness of algorithm 1.3 for identifying palindromes.

Problem 4.32. *Show that $L_{pal} = L(G_{pal})$, in other words, show that G_{pal} is the grammar that generates precisely the set of all palindromes over $\{0,1\}$. To show $\subseteq$ use induction on the length of strings in order to show that each palindrome has a derivation; for $\supseteq$ use induction on the length of derivations.*

A context-free grammar is in *Chomsky Normal Form (CNF)* if all the rules are of the form: $A \longrightarrow BC$, $A \longrightarrow a$, and $S \longrightarrow \varepsilon$, where A, B, C are variables (not necessarily distinct), a is a terminal, and S is the starting variable (so note that ε can only be generated from the starting variable).

The grammar G_{pal} can be translated into CNF as follows: first the rule $S \longrightarrow \varepsilon|0|1$ is allowed as is, so we only have to deal with $S \longrightarrow 0S0$ and $S \longrightarrow 1S1$. To do that, we introduce two new variables U_0, U_1 and the following two rules, $U_0 \longrightarrow 0$ and $U_1 \longrightarrow 1$, and replace $S \longrightarrow 0S0|1S1$ with $S \longrightarrow U_0SU_0|U_1SU_1$. Now the only problem is that there are too many variables (we are only allowed 2 per body of rule in CNF), so we introduce two more new variables V_0, V_1 and transform $S \longrightarrow U_0SU_0|U_1SU_1$ as follows: $S \longrightarrow U_0V_0$, $V_0 \longrightarrow SU_0$, and $S \longrightarrow U_1V_1$, $V_1 \longrightarrow SU_1$. Putting it all together, we have a new grammar G'_{pal} given by $(\{S, U_0, U_1, V_0, V_1\}, \{0,1\}, R, S)$ and R is given by the following rules:

$$\{S \longrightarrow \varepsilon|0|1, U_0V_0, U_1V_1, U_0 \longrightarrow 0, U_1 \longrightarrow 1, V_0 \longrightarrow SU_0, V_1 \longrightarrow SU_1\},$$

and note that G'_{pal} is also a correct grammar for the language of palindromes, and furthermore it is in CNF.

Problem 4.33. *Every context-free grammar can be transformed to an equivalent grammar (i.e., generating the same set of strings of terminals) in CNF. (Hint: the main ideas are given in the above example; you will have to deal with the case $X \longrightarrow Y$ (the so called* unit rules*), and $X \longrightarrow \varepsilon$ where X is not the starting variable (the so called* ε-rules*).)*

Given a grammar G in CNF, and a string $w = a_1a_2 \ldots a_n$, we can test whether $w \in L(G)$ using the CYK[4] dynamic algorithm (algorithm 4.6). On input $G, w = a_1a_2 \ldots a_n$ algorithm 4.6 builds an $n \times n$ table T, where each entry contains a subset of V. At the end, $w \in L(G)$ iff the start variable S is contained in position $(1, n)$ of T. The main idea is to put variable X_1 in position (i, j) if X_2 is in position (i, k) and X_3 is in position $(k+1, j)$ and $X_1 \longrightarrow X_2X_3$ is a rule. The reasoning is that X_1 is in position (i, k) iff $X_1 \overset{*}{\Rightarrow} a_i \ldots a_k$, that is, the substring $a_i \ldots a_k$ of the input string can be generated from X_1. Let $V = \{X_1, X_2, \ldots, X_m\}$.

[4] Named after the inventors: Cocke-Younger-Kasami.

Algorithm 4.6 CYK

for $i = 1..n$ **do**
  for $j = 1..m$ **do**
    Place variable X_j in (i,i) iff $X_j \longrightarrow a_i$ is a rule of G
  end for
end for
for $1 \leq i < j \leq n$ **do**
  for $k = i..(j-1)$ **do**
    if $(\exists X_p \in (i,k) \wedge \exists X_q \in (k+1,j) \wedge \exists X_r \longrightarrow X_pX_q)$ **then**
      Put X_r in (i,j)
    end if
  end for
end for

×	(2,2)	(2,3)	(2,4)	(2,5)
×	×			(3,5)
×	×	×		(4,5)
×	×	×	×	(5,5)

Fig. 4.5 Computing the entry $(2,5)$: note that we require all the entries on the same row and column (except those that are below the main diagonal). Thus the CYK algorithm computes the entries dynamically by diagonals, starting with the main diagonal, and ending in the upper-right corner.

In the example in figure 4.5, we show which entries in the table we need to use to compute the contents of $(2,5)$.

Problem 4.34. *Show the correctness of algorithm 4.6.*

Problem 4.35. *Implement the CYK algorithm in Python. Choose a convention for representing CFGs, and document it well in your code. You may assume the grammar is given in CNF; or, you may check that explicitly. To make the project even more ambitious, you may implement a translation of a general grammar to CNF, following your solution to problem 4.33.*

Context free grammars are the foundations of parsers. There are many tools that implement the ideas mentioned in this section; for example, `Lex`, `Yacc`, `Flex`, `Bison`, and others. you may read more about them here: `http://dinosaur.compilertools.net`.

4.7 Answers to selected problems

Problem 4.5. Consider the graph G_n in figure 4.6. There are $2+n+n=2n+2$ nodes in it, and there are 2^n paths from s to t. To see that, note that starting at s we have a choice to go to node 1 or node $1'$. And then we always have a choice to go up or down, so $2 \times 2^{n-1}$ paths that land us at n or n', and finally we just go to t. Note that we have given an undirected graph; but simply giving all the edges a "left-to-right" direction gives us an example for directed graphs.

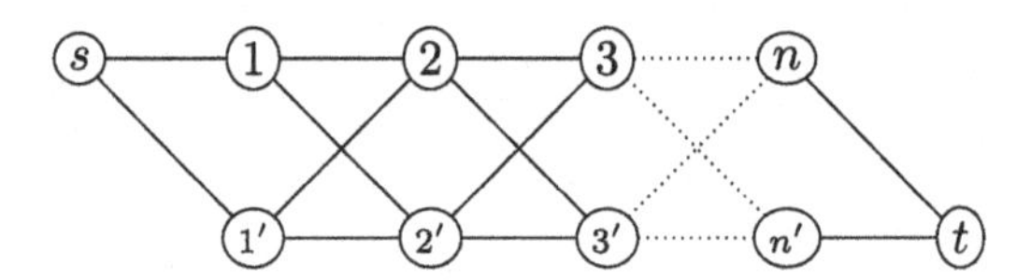

Fig. 4.6 Exponentially many paths (problem 4.5).

Problems 4.6 and 4.7. The pre-condition is that $\forall i,j \in [n]$ we have that $B(i,j)=A(0,i,j)$. The post-condition is that $\forall i,j \in [n]$ we have that $B(i,j)=A(n,i,j)$. The loop invariant is that after the k-th iteration of the main loop, $B(i,j)=A(k,i,j)$. To prove the loop invariant note that $B(i,j)$ is given by $\min\{B(i,j), B(i,k)+B(k,j)\}$, so the only worry is that $B(i,k)$ or $B(k,j)$ was *already* updated, so we are not getting $A(k-1,i,k)$ or $A(k-1,k,j)$ as we should, but rather $A(k,i,k)$ or $A(k,k,j)$. But, it turns out that $A(k,i,k)=A(k-1,i,k)$ and $A(k,k,j)=A(k-1,k,j)$, because the shortest path from i to k (or k to j) does *not* contain k as an intermediate node.

Problem 4.10. Overwriting does not create a problem, because the values of j are considered in decreasing order $C, C-1, \ldots, 1$. Thus the array position $S(j-w_i)$ has not yet been updated when the reference is made.

Problem 4.11. The pre-condition is that for all j, $S(j)=R(0,j)$. The post-condition is that for all j, $S(j)=R(d,j)$. Let the loop invariant be the assertion that after the i-th step, $S(j)=R(i,j)$. This loop invariant holds since we start "filling" S in from the right, and we only change false to true (never true to false—the reason is that if we could build an intermediate value j with the first $(i-1)$ weights, we can certainly still construct it with the first i weights).

Problem 4.12. Consider $w_1=w_2=1$, $w_3=2$ and $C=3$, so the table on this input would look as follows:

	0	1	2	3
	T	F	F	F
$w_1 = 1$	T	T	F	F
$w_2 = 1$	T	T	T	F
$w_3 = 2$	T	T	T	T

Now consider the row for $w_1 = 1$, and the entry for the column labeled with 2. That entry is an F, as it should be, but if the for-loop in algorithm 4.3 were not a decreasing loop, then we would update that entry to a T since for $j = 2$, we have that $2 \geq w_1$ and $S(2 - w_1) = \mathsf{T}$.

Problem 4.14. First, we need to find the solution; so we look in the last row (i.e., row d) for the largest non-zero j. That is, the solution is given by $M = \max_{0 \leq j \leq C}[R(d, j) = \mathsf{T}]$. Now we check if $R(d-1, M) = \mathsf{T}$. If yes, then we know that weight w_d is not necessary, so we do not include it, and continue looking at $R(d-2, M)$. If no, then because $R(d, M) = \mathsf{T}$, we know that $M - w_d \geq 0 \wedge R(d-1, M - w_d) = \mathsf{T}$. So we include w_d, and continue looking at $R(d-1, M - w_d)$. We stop when we reach the first column of the array.

Problem 4.15. The natural greedy algorithm that attempts to solve SKS is the following: order the weights from the heaviest to the lightest, and add them in that order for as long as possible. Assume that $\overline{M} \neq M$, and let S_0 be the result of this greedy procedure, i.e., a subset of $\{1, \ldots, d\}$ such that $K(S_0) = \overline{M}$. First show that there is at least one weight in S_0: If $S_0 = \emptyset$, then $\overline{M} = 0$, and all weights must be larger than C, but then $M = 0$, so $\overline{M} = M$ which is not the case by assumption. Now show that there is at least one weight not in S_0: if all the weights are in S, then again $\overline{M} = \sum_{i=1}^{d} w_i = M$. Finally, show now that $\overline{M} > \frac{1}{2}C$ by considering the first weight, call it w_j, which has been rejected after at least one weight has been added (note that such a weight must exist; we may assume that there are no weights larger than the capacity C, and if there are we can just not consider them; therefore, the first weight on the list is added, and then we know that some weight will come along which won't be added; we consider the first such weight): If $w_j \leq \frac{1}{2}C$, then the sum of the weights which are already in is $> \frac{1}{2}C$, so $\overline{M} > \frac{1}{2}C$. If $w_j > \frac{1}{2}C$, then, since the objects are ordered by greedy in non-increasing order of weights, the weights that are already in are $> \frac{1}{2}C$, so again $\overline{M} > \frac{1}{2}C$.

Problem 4.16. In the first space put $w_i + \sum_{j \in S} w_j \leq C$ and in the second space put $S \longleftarrow S \cup \{i\}$.

Problem 4.17. Define "S is promising" to mean that S can be extended, using weights which have *not* been considered yet, to an optimal solutions $S_{\max}$. At the end, when no more weights have been left to consider, the loop invariant still holds true, so S itself must be optimal.

We show that "S is promising" is a loop invariant by induction on the number of iterations. Basis case: $S = \emptyset$, so S is clearly promising. Induction step: Suppose that S is promising (so S can be extended, using weights which have not been considered yet, to $S_{\max}$). Let S' be S after one more iteration. Suppose $i \in S'$. Since $w_i \geq \sum_{j=i+1}^{d} w_j$, it follows that:

$$K(S') \geq K(S) + \sum_{j=i+1}^{d} w_j \geq K(S_{\max})$$

so S' is already maximal, and we are done. Suppose $i \notin S'$; then we have that $w_i + \sum_{j \in S} w_j > C$, so $i \notin S_{\max}$, so S' can be extended (using weights which have not been considered yet!) to $S_{\max}$. In either case, S' is promising.

Problem 4.18. $V(i,j) = 0$ if $i = 0$ or $j = 0$. And for $i, j > 0$, $V(i,j)$ is

$$\begin{cases} V(i-1,j) & \text{if } j < w_i \text{ or } R(i-1, j-w_i) = \mathsf{F} \\ \max\{v_i + V(i-1, j-w_i), V(i-1,j)\} & \text{otherwise} \end{cases}$$

To see that this works, suppose that $j < w_i$. Then weight i cannot be included, so $V(i,j) = V(i-1,j)$. If $R(i-1, j-w_i) = \mathsf{F}$, then there is no subset $S \subseteq \{1, \ldots, i\}$ such that $i \in S$ and $K(S) = j$, so again weight i is not included, and $V(i,j) = V(i-1,j)$.

Otherwise, if $j \geq w_i$ and $R(i-1, j-w_i) = \mathsf{T}$, then weight i may or may not be included in S. We take the case which offers more value: $\max\{v_i + V(i-1, j-w_i), V(i-1,j)\}$.

Problem 4.19. By changing the definition of $V(i,j)$ given in (4.2) to have $K(S) \leq j$ (instead of $K(S) = j$), we can take the recurrence given for V in the solution to problem 4.18 and simply get rid of the part "or $R(i-1, j-w_i) = \mathsf{F}$" to obtain a recurrence for V that does not require computing R.

Problem 4.21. The algorithm must include a computation of the distinct finish times, i.e., the u_i's, as well as a computation of the array H. Here we just give the algorithm for computing A based on the recurrence (4.3). The assumption is that there are n activities and k distinct finish times.

Problem 4.22. We show how to find the actual set of activities: Suppose $k > 0$. If $A(k) = A(k-1)$, then no activity has been scheduled to end at time u_k, so we proceed recursively to examine $A(k-1)$. If, on the other hand, $A(k) \neq A(k-1)$, then we know that some activity has been

Algorithm 4.7 Activity selection

```
A(0) ⟵ 0
for j : 1..k do
        max ⟵ 0
        for i = 1..n do
                if f_i = u_j then
                        if p_i + A(H(i)) > max then
                                max ⟵ p_i + A(H(i))
                        end if
                end if
        end for
        if A(j − 1) > max then
                max ⟵ A(j − 1)
        end if
        A(j) ⟵ max
end for
```

scheduled to end at time u_k. We have to find out which one it is. We know that in this case $A(k) = \max_{1 \le i \le n}\{p_i + A(H(i)) | f_i = u_k\}$, so we examine all activities i, $1 \le i \le n$, and output the (first) activity i_0 such that $A(k) = p_{i_0} + A(H(i_0))$ and $f_{i_0} \le u_k$. Now we repeat the procedure with $A(H(i_0))$. We end when $k = 0$.

Problem 4.24. Initialization: $A(0,t) = 0$, $0 \le t \le t_n$. To compute $A(i,t)$ for $i > 0$ first define $t_{\min} = \min\{t, t_i\}$. Now

$$A(i,t) = \begin{cases} A(i-1,t) & \text{if } t_{\min} < d_i \\ \max\{A(i-1,t), p_i + A(i-1, t_{\min} - d_i)\} & \text{otherwise} \end{cases}$$

Justification: If job i is scheduled in the optimal schedule, it finishes at time $t_{\min}$, and starts at time $t_{\min} - d_i$. If it is scheduled, the maximum possible profit is $A(i-1, t_{\min} - d_i) + p_i$. Otherwise, the maximum profit is $A(i-1,t)$.

Problem 4.26. $M = \max_{1 \le j \le n} M(j)$

Problem 4.27.

(1) $r_1 (= S_{11})$
(2) $M(j-1) > 0$
(3) $M(j-1) + r_j$
(4) r_j

4.8 Notes

Any algorithms textbook will have a section on dynamic programming; see for example chapter 15 in [Cormen *et al.* (2009)] and chapter 6 in [Kleinberg and Tardos (2006)].

While matroids serve as a good abstract model for greedy algorithms, a general model for dynamic programming is being currently developed. See [Michael Aleknovich (2005)].

Chapter 5

Online Algorithms

The algorithms presented in the previous chapters were *offline* algorithms, in the sense that the entire input was given at the beginning. In this chapter we change our paradigm and consider *online* algorithms, where the input is presented piecemeal, in a serial fashion, and the algorithm has to make decisions based on incomplete information, without knowledge of future events.

A typical example of an application is a caching discipline; consider a hard disk from which data is read into a random access memory. Typically, the random access memory is much smaller, and so it must be decided which data has to be overwritten with new data. New requests for data from the hard disk arrive continuously, and it is hard to predict future requests.

Thus we must overwrite parts of the random access memory with new requests, but we must do the overwriting judiciously, so that we minimize future *misses*: data that is required but not present in the random access memory, and so it has to be brought in from the hard disk. Minimizing the number of misses is difficult when the future requests are unknown.

Correctness in the context of online algorithm has a different nuance; it means that the algorithm minimizes strategic errors. That is, an online algorithm will typically do worse than a corresponding offline algorithm that sees the entire input, but we want it to be as competitive as possible, given its intrinsic limitations. Thus, in the context of online algorithms, we are concerned with performance evaluation.

We introduce the subject of online algorithms with the list accessing problem in section 5.1, and then present paging algorithms in section 5.2.

5.1 List accessing problem

We are in charge of a filing cabinet containing l labeled but unsorted files. We receive a sequence of requests to access files; each request is a file label. After receiving a request for a file we must locate it, process it, and return it to the cabinet.

Since the files are unordered we must flip through the files starting at the beginning, until the requested file is located. If a file is in position i, we incur a search cost of i in locating it. If the file is not in the cabinet, the cost is l, which is the total number of files. After taking out the file, we must return it to the cabinet, but we may choose to reorganize the cabinet; for instance, we might put it closer to the front. The incentive for such a reorganization is that it may save us some search time in the future: if a certain file is requested frequently, it is wise to insert it closer to the front. Our goal is to find a reorganization rule that minimizes the search time.

Let $\sigma = \sigma_1, \sigma_2, \ldots, \sigma_n$ be a finite sequence of n requests. To service request σ_i, a list accessing algorithm ALG must search for the item labeled σ_i by traversing the list from the beginning, until it finds it. The cost of retrieving this item is the index of its position on the list. Thus, if item σ_i is in position j, the cost of retrieving it is j. Furthermore, the algorithm may reorganize the list at any time.

The work associated with a reorganization is the minimum number of transpositions of consecutive items needed to carry it out. Each transposition has a cost of 1, however, immediately after accessing an item, we allow it to be moved free of charge to any location closer to the front of this list. These are *free* transpositions, while all other transpositions are *paid*. Let ALG(σ) be the sum of the costs of servicing all the items on the list σ, i.e., the sum of the costs of all the searches plus the sum of the costs of all paid transpositions.

Problem 5.1. *What is the justification for this "free move"? In other words, why does it make sense to allow placing an item "for free" right after accessing it? Finally, show that given a list of l items, we can always reorder it in any way we please by doing only transpositions of consecutive items.*

We consider the *static list accessing model*, where we have a list of l items, and the only requests are to access an item on the list, i.e., there are no insertions or deletions. Many algorithms have been proposed for managing lists; we are going to examine *Move To Front (MTF)*, where after accessing an item, we move it to the front of the list, without changing the relative order of the other items.

Further, we assume that σ consists of *only* those items which appear on the list of MTF—this is not a crucial simplification; see problem 5.7. Notice that MTF(σ) is simply the sum of the costs of all the searches, since we only change the position of an item when we retrieve it, in which case we move it for free to the front.

Theorem 5.2. *Let* OPT *be an optimal (offline) algorithm for the static list accessing model. Suppose that* OPT *and* MTF *both start with the same list configuration. Then, for any sequence of requests* σ*, where* $|\sigma| = n$*, we have that*

$$\text{MTF}(\sigma) \leq 2 \cdot \text{OPT}_S(\sigma) + \text{OPT}_P(\sigma) - \text{OPT}_F(\sigma) - n, \tag{5.1}$$

where $\text{OPT}_S(\sigma), \text{OPT}_P(\sigma), \text{OPT}_F(\sigma)$ *are the total cost of searches, the total number of paid transpositions and the total number of free transpositions, of* OPT *on* σ*, respectively.*

Proof. Imagine that both MTF and OPT process the requests in σ, while each algorithm works on its own list, starting from the same initial configuration. You may think of MTF and OPT as working in parallel, starting from the same list, and neither starts to process σ_i until the other is ready to do so.

Let

$$a_i = t_i + (\Phi_i - \Phi_{i-1}) \tag{5.2}$$

where t_i is the actual cost that MTF incurs for processing this request (so t_i is in effect the position of item σ_i on the list of MTF *after* the first $i-1$ requests have been serviced). Φ_i is a *potential function*, and here it is defined as the number of *inversions* in MTF's list with respect to OPT's list. An inversion is defined to be an ordered pair of items x_j and x_k, where x_j precedes x_k in MTF's list, but x_k precedes x_j in OPT's list.

Problem 5.3. *Suppose that* $l = 3$*, and the list of* MTF *is* x_1, x_2, x_3*, and the list of* OPT *is* x_3, x_2, x_1*. What is* Φ *in this case? In fact, how can we compute* $\text{OPT}(\sigma)$*, where* σ *is an arbitrary sequence of requests, without knowing how* OPT *works?*

Note that Φ_0 depends only on the initial configurations of MTF and OPT, and since we assume that the lists are initially identical, $\Phi_0 = 0$. Finally, the value a_i in (5.2) is called the *amortized cost*, and its intended meaning is the cost of accessing σ_i, i.e., t_i, plus a measure of the increase of the "distance" between MTF's list and OPT's list after processing σ_i, i.e., $\Phi_i - \Phi_{i-1}$.

It is obvious that the cost incurred by MTF in servicing σ, denoted MTF(σ), is $\sum_{i=1}^n t_i$. But instead of computing $\sum_{i=1}^n t_i$, which is difficult, we compute $\sum_{i=1}^n a_i$ which is much easier. The relationship between the two summations is,

$$\text{MTF}(\sigma) = \sum_{i=1}^{n} t_i = \Phi_0 - \Phi_n + \sum_{i=1}^{n} a_i, \tag{5.3}$$

and since we agreed that $\Phi_0 = 0$, and Φ_i is always positive, we have that,

$$\text{MTF}(\sigma) \leq \sum_{i=1}^{n} a_i. \tag{5.4}$$

So now it remains to compute an upper bound for a_i.

Problem 5.4. *Show the second equality of equation (5.3).*

Assume that the i-th request, σ_i, is in position j of OPT, and in position k of MTF (i.e., this is the position of this item *after* the first $(i-1)$ requests have been completed). Let x denote this item—see figure 5.1.

We are going to show that

$$a_i \leq (2s_i - 1) + p_i - f_i, \tag{5.5}$$

where s_i is the search cost incurred by OPT for accessing request σ_i, and p_i and f_i are the paid and free transpositions, respectively, incurred by OPT when servicing σ_i. This shows that

$$\begin{aligned}\sum_{i=1}^{n} a_i &\leq \sum_{i=1}^{n}((2s_i - 1) + p_i - f_i) \\ &= 2(\sum_{i=1}^{n} s_i) + (\sum_{i=1}^{n} p_i) - (\sum_{i=1}^{n} f_i) - n \\ &= 2\text{OPT}_S(\sigma) + \text{OPT}_P(\sigma) - \text{OPT}_F(\sigma) - n,\end{aligned}$$

which, together with the inequality (5.4), will show (5.1).

We prove (5.5) in two steps: in the first step MTF makes its move, i.e., moves x from the k-th slot to the beginning of its list, and we measure the change in the potential function *with respect to* the configuration of the list of OPT *before* OPT makes its own moves to deal with the request for x.

In the second step, OPT makes its move and now we measure the change in the potential function *with respect to* the configuration of the list of MTF

MTF	*	*		*		x			
OPT				x		*	*	*	

Fig. 5.1 x is in position k in MTF, and in position j in OPT. Note that in the figure it appears that $j < k$, but we make no such assumption in the analysis. Let $*$ denote items located before x in MTF but after x in OPT, i.e., the $*$ indicate inversions with respect to x. There may be other inversions involving x, namely items which are after x in MTF but before x in OPT, but we are not concerned with them.

after MTF has completed its handling of the request (i.e., with x at the beginning of the list of MTF).

See figure 5.1: suppose that there are v such $*$, i.e., v inversions of the type represented in the figure. Then, there are at least $(k-1-v)$ items that precede x in both list.

Problem 5.5. *Explain why at least $(k-1-v)$ items precede x in both lists.*

But this implies that $(k-1-v) \leq (j-1)$, since x is in the j-th position in OPT. Thus, $(k-v) \leq j$. So what happens when MTF moves x to the front of the list? In terms of inversions two things happen: (i) $(k-1-v)$ new inversions are created, with respect to OPT's list, before OPT itself deals with the request for x. Also, (ii) v inversions are eliminated, again with respect to OPT's list, before OPT itself deals with the request for x.

Therefore, the contribution to the amortized cost is:

$$k + ((k-1-v) - v) = 2(k-v) - 1 \overset{(1)}{\leq} 2j - 1 \overset{(2)}{=} 2s - 1 \tag{5.6}$$

where (1) follows from $(k-v) \leq j$ shown above, and (2) follows from the fact that the search cost incurred by OPT when looking for x is exactly j. Note that (5.6) looks similar to (5.5), but we are missing $+p_i - f_i$. These terms will come from considering the second step of the analysis: OPT makes its move and we measure the change of potential with respect to MTF with x at the beginning of the list. This is dealt with in the next problem.

Problem 5.6. *In the second step of the analysis,* MTF *has made its move and* OPT*, after retrieving x, rearranges its list. Show that each paid transposition contributes 1 to the amortized cost and each free transposition contributes -1 to the amortized cost.*

This finishes the proof. □

In the *dynamic list accessing model* we also have *insertions*, where the cost of an insertion is $l+1$—here l is the length of the list—, and *deletions*, where the cost of a deletion is the same as the cost of an access, i.e., the position of the item on the list.

Problem 5.7. *Show that theorem 5.2 still holds in the dynamic case.*

The *infimum* of a subset $S \subseteq \mathbb{R}$ is the largest element r, not necessarily in S, such that for all all $s \in S$, $r \leq s$. We say that an online algorithm is *c-competitive* if there is a constant α such that for all finite input sequences $\mathrm{ALG}(\sigma) \leq c \cdot \mathrm{OPT}(\sigma) + \alpha$. The infimum over the set of all values c such that ALG is c-competitive is called the *competitive ratio* of ALG and is denoted $\mathcal{R}(\mathrm{ALG})$.

Problem 5.8. *Observe that* $\mathrm{OPT}(\sigma) \leq n \cdot l$, *where* l *is the length of the list and* n *is* $|\sigma|$.

Problem 5.9. *Show that* MTF *is a 2-competitive algorithm, and that* $\mathcal{R}(\mathrm{MTF}) \leq 2 - \frac{1}{l}$.

Problem 5.10. *In the chapters on online and randomized algorithms (this chapter and the next) we need to generate random values. Use the Python* `random` *library to generate those random values; implement* OPT *and* MTF *and compare them on a random sequence of requests. You may want to plot the competitiveness of* MTF *with respect to* OPT *using* `gnuplot`.

The traditional approach to studying online algorithms falls within the framework of *distributional*, also known as *average-case*, complexity: a distribution on event sequences is hypothesizes, and the expected payoff per event is analyzed. However, in this chapter we present a more recent approach to *competitive analysis*, whereby the payoff of an online algorithm is measured by comparing its performance to that of an *optimal offline algorithm.* Competitive analysis thus falls within the framework of *worst case* complexity.

5.2 Paging

Consider a two-level *virtual memory system*: each level, slow and fast, can store a number of fixed-size memory units called *pages*. The slow memory stores N pages, and the fast memory stores k pages, where $k < N$. The k is usually much smaller than N.

Given a request for page p_i, the system must make page p_i available in the fast memory. If p_i is already in the fast memory, called a *hit*, the system need not do anything. Otherwise, on a *miss*, the system incurs a *page fault*, and must copy the page p_i from the slow memory to the fast memory. In doing so, the system is faced with the following problem: which page to evict from the fast memory to make space for p_i. In order to *minimize* the number of page faults, the choice of which page to evict must be made wisely.

Typical examples of fast and slow memory pair are a RAM and hard disk, respectively, or a processor-cache and RAM, respectively. In general, we shall refer to the fast memory as "the cache." Because of its important role in the performance of almost every computer system, paging has been extensively studied since the 1960s, and the common paging schemes are listed in figure 5.2.

All the caching disciplines in figure 5.2, except for the last one, are online algorithms; that is, they are algorithms that make decisions based on past events, rather than the future. The last algorithm, LFD, replaces the page whose next request is the latest, which requires knowledge of future requests, and hence it is an offline algorithm.

5.2.1 *Demand paging*

Demand paging algorithms never evict a page from the cache unless there is a page fault. All the paging disciplines in figure 5.2 are demand paging.

We consider the *page fault model*: in this somewhat simplistic model, we charge 1 for bringing a page into the fast memory, and we charge 0 for accessing a page which is already there—in practice there are other costs involved. As the next theorem shows we lose nothing by restricting our attention to this model.

LRU	*Least Recently Used*
CLOCK	*Clock Replacement*
FIFO	*First-In/First-Out*
LIFO	*Last-In/First-Out*
LFU	*Least Frequently Used*
LFD	*Longest Forward Distance*

Fig. 5.2 Paging disciplines: the top five are online algorithms; the last one, LFD, is an offline algorithm. We shall see in section 5.2.6 that LFD is in fact the optimal algorithm for paging.

Theorem 5.11. *Any page replacement algorithm, online or offline, can be modified to be demand paging without increasing the overall cost on any request sequence.*

Proof. In a demand paging algorithm a page fault causes exactly one eviction (once the cache is full, that is), and there are no evictions between misses. So let ALG be any paging algorithm. We show how to modify it to make it a demand paging algorithm ALG′, in such a way that on any input sequence ALG′ incurs at most the cost (makes at most as many page moves from slow to fast memory) as ALG, i.e., $\forall\sigma$, $\text{ALG}'(\sigma) \leq \text{ALG}(\sigma)$.

Suppose that ALG has a cache of size k. Define ALG′ as follows: ALG′ also has a cache of size k, plus k registers. ALG′ runs a simulation of ALG, keeping in its k registers the page numbers of the pages that ALG would have had in its cache. Based on the behavior of ALG, ALG′ makes decisions to evict pages[1].

Suppose page p is requested. If p is in the cache of ALG′, then just service the request. Otherwise, if a page fault occurs, ALG′ behaves according to the following two cases:

Case 1. If ALG also has a page fault (that is, the number of p is *not* in the registers), and ALG evicts a page from register i to make room for p, then ALG′ evicts a page from slot i in its cache, to make room for p.

Case 2. If ALG does not have a page fault, then the number of p must be in, say, register i. In that case, ALG′ evicts the contents of slot i in its cache, and moves p in there.

Thus ALG′ is a demand paging algorithm.

We now show that ALG′ incurs at most the cost of ALG on any input sequence; that is, ALG′ has at most as many page faults as ALG. To do this, we pair each page move of ALG′ with a page move of ALG in a unique manner as follows: If ALG′ and ALG both incur a page fault, then match the corresponding page moves. Otherwise, if ALG already had the page in its cache, it must have moved it there before, so match that move with the current move of ALG′.

It is never the case that two different moves of ALG′ are matched with a single move of ALG. To see this, suppose that on some input sequence, we encounter for the first time the situation where two moves of ALG′ are matched with the same move of ALG. This can only happen in the following

[1]The assumption in this proof is that ALG does not re-arrange its slots—i.e., it never permutes the contents of its cache.

situation: page p is requested, ALG$'$ incurs a page fault, it moves p into its cache, and we match this move with a past move of ALG, which has been matched already! But this means that page p was already requested, and after it has been requested, it has been evicted from the cache of ALG$'$ (otherwise, ALG$'$ would not have had a page fault).

Thus, ALG$'$ evicted page p while ALG did not, so they were not in the same slot. But ALG$'$ put (the first time) p in the same slot as ALG! Contradiction. Therefore, we could not have matched a move twice. Thus, we can match each move of ALG$'$ with a move of ALG, in a one-to-one manner, and hence ALG$'$ makes at most as many moves as ALG. See figure 5.3.

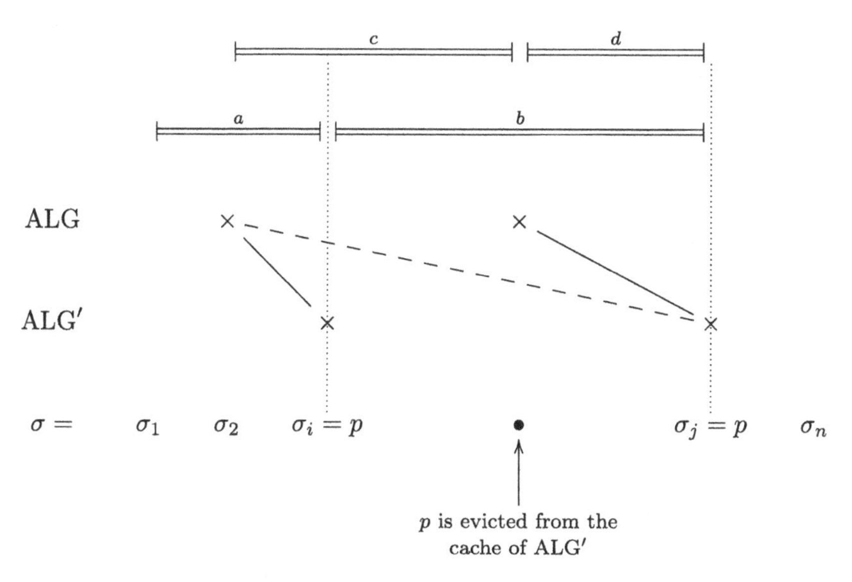

Fig. 5.3 Suppose that i, j is the smallest pair such that there exists a page p with the property that $\sigma_i = \sigma_j = p$ and ALG$'$ incurs a page fault at σ_i and σ_j and the two corresponding page moves of ALG$'$ are both matched with the same page move of p by ALG somewhere in the stretch a. We show that this is not possible: if ALG$'$ incurs a page fault at $\sigma_i = \sigma_j = p$ it means that somewhere in b the page p is evicted—this point is denoted with '•'. If ALG did not evict p in the stretch c, then ALG also evicts page p at '•' and so it must then bring it back to the cache in stretch d—we would match the $\times$ at σ_j with that move. If ALG did evict p in the stretch c, then again it would have to bring it back in before σ_j. In any case, there is a later move of p that would be matched with the page fault of ALG$'$ at σ_j.

Problem 5.12. *In figure 5.3 we postulate the existence of a "smallest" pair i, j with the given properties. Show that if such a pair exists then there exists a "smallest" such pair; what does "smallest" mean in this case?*

The idea is that ALG does not gain anything by moving a page into its cache preemptively (before the page is actually needed). ALG′ waits for the request before taking the same action.

In the meantime (between the time that ALG moves in the page and the time that it is requested and ALG′ brings it in), ALG′ can only gain, because there are no requests for that page during that time, but there might be a request for the page that ALG evicted preemptively.

Note that in the simulation, ALG′ only needs k extra registers, to keep track of the page numbers of the pages in the cache of ALG, so it is an *efficient* simulation. □

Theorem 5.11 allows for us to restrict our attention to demand paging algorithms, and thus use the terms "page faults" and "page moves" interchangeably, in the sense that in the context of demand paging, we have a page move if and only if we have a page fault.

5.2.2 *FIFO*

When a page must be replaced, the *oldest* page is chosen. It is not necessary to record the time when a page was brought in; all we need to do is create a FIFO (First-In/First-Out) queue to hold all pages in memory. The FIFO algorithm is easy to understand and program, but its performance is not good in general.

FIFO also suffers from the so called *Belady's anomaly.* Suppose that we have the following sequence of page requests: $1, 2, 3, 4, 1, 2, 5, 1, 2, 3, 4, 5$. Then, we have more page faults when $k = 4$ than when $k = 3$. That is, FIFO has more page faults with a bigger cache!

Problem 5.13. *For a general i, provide a sequence of page requests that illustrates Belady's anomaly incurred by* FIFO *on cache sizes i and $i + 1$. In your analysis, assume that the cache is initially empty.*

5.2.3 *LRU*

The optimal algorithm for page replacement, OPT, evicts the page whose next request is the latest, and if some pages are never requested again, then anyone of them is evicted. This is an impractical algorithm from the point of view of online algorithms as we do not know the future.

However, if we use the recent past as an approximation of the near future, then we will replace the page that *has not been used for the longest period of time.* This approach is the *Least Recently Used (LRU)* algorithm.

LRU replacement associates with each page the time of that page's last use. When a page must be replaced, LRU chooses that page that has not been used for the longest period of time. The LRU algorithm is considered to be good, and is often implemented—the major problem is *how* to implement it; two typical solutions are counters and stacks.

Counters: Keep track of the time when a given page was last referenced, updating the counter every time we request it. This scheme requires a search of the page table to find the LRU page, and a write to memory for each request; an obvious problem might be clock overflow.

Stack: Keep a stack of page numbers. Whenever a page is referenced, it is removed from the stack and put on the top. In this way, the top of the stack is always the most recently used page, and the bottom is the LRU page. Because entries are removed from the middle of the stack, it is best implemented by a doubly-linked list.

How many pointer operations need to be performed in the example in figure 5.4? Six, if we count as follows: remove old head and add new head (2 operations), connect 4 with 1 (2 operations), connect 3 with 5 (2 operations). However, we could have also counted disconnecting 3 with 4 and 4 with 5, giving 4 more pointer operations, giving us a total of 10. A third strategy would be not to count disconnecting pointers, in which case we would get half of these operations, 5. It does not really matter how we count, because the point is that in order to move a requested page (after a hit) to the top, we require a small *constant* number of pointer operations, regardless of how we count them.

Problem 5.14. *List the pointer operations that have to be performed if the requested page is* not *in the cache. Note that you should list the pointer operations (not just give a "magic number"), since we just showed that there are three different (all reasonable) ways to count them. Again, the point is, that if a page has to be brought in from the slow memory to the cache, a small constant number of pointer operations have to be performed.*

Problem 5.15. *We have implemented* LRU *with a doubly-linked list. What would be the problem if we used a normal linked list instead? That is, if every page had only a pointer to the next page: $i \rightsquigarrow j$, meaning that i was*

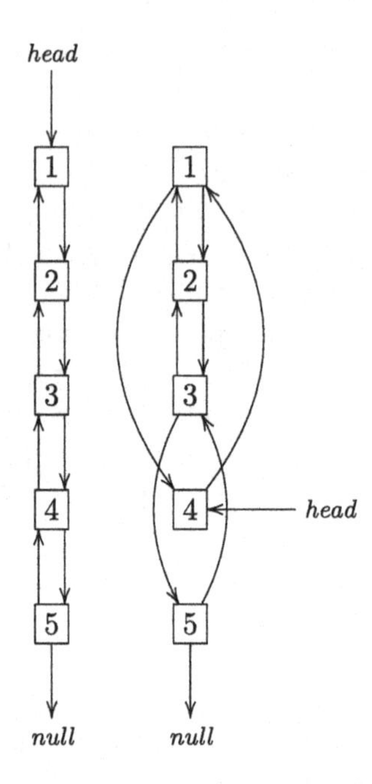

Fig. 5.4 LRU stack implementation with a doubly-linked list. The requested page is page 4; the left list shows the state *before* page 4 is requested, and the right list shows the state *after* the request has been serviced.

requested more recently than j, but no page was requested later than i and sooner than j.

Lemma 5.16. LRU *does* not *incur Belady's anomaly (on any cache size and any request sequence).*

Proof. Let $\sigma = p_1, p_2, \ldots, p_n$ be a request sequence, and let $\text{LRU}_i(\sigma)$ be the number of faults that LRU incurs on σ with a cache of size i. We show that for all i and σ, the following property holds:

$$\text{LRU}_i(\sigma) \geq \text{LRU}_{i+1}(\sigma). \tag{5.7}$$

Once we show (5.7), it follows that for any pair $i < j$ and any request sequence σ, $\text{LRU}_i(\sigma) \geq \text{LRU}_j(\sigma)$, and conclude that LRU does not incur Belady's anomaly.

To show (5.7), we define a property of doubly-linked lists which we call "embedding." We say that a doubly-linked list of size i can be *embedded* in another doubly-linked list of size $i + 1$, if the two doubly-linked lists are identical, except that the longer one may have one more item at the

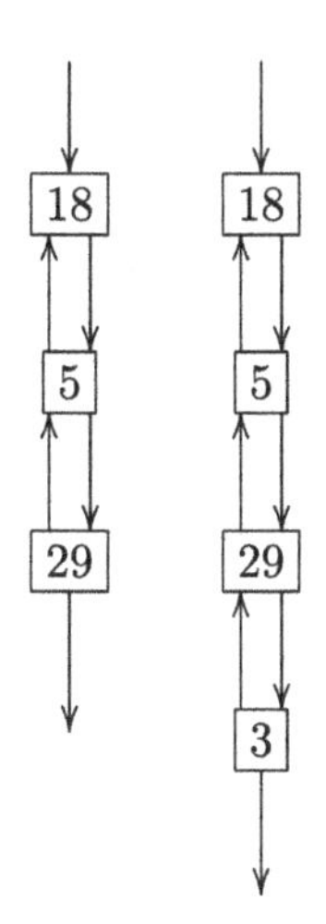

Fig. 5.5 The list on the left can be *embedded* into the list on the right.

"end." See figure 5.5, where the doubly-linked list of size 3 on the left can be embedded in the doubly-linked list of size 4 on the right.

At the beginning of processing the request sequence, when the caches are getting filled up, the two lists are identical, but once the caches are full, the LRU_{i+1} cache will have one more item.

Claim 5.17. *After processing each request, the doubly-linked list of* LRU_i *can be embedded into the doubly-linked list of* LRU_{i+1}.

Proof. We prove this claim by induction on the number of steps. Basis case: if $n = 1$, then both LRU_i and LRU_{i+1} incur a fault and bring in p_1. Induction step: suppose that the claim holds after step n; we show that it also holds after step $n + 1$. Consider the following cases: (1) LRU_i has a hit on p_{n+1}, (2) LRU_i has a fault on p_{n+1}, (2a) LRU_{i+1} also has a fault, (2b) LRU_{i+1} does not have a fault.

Problem 5.18. *Show that in each case the embedding property is being preserved.*

This finishes the proof of the claim. □

Problem 5.19. *Use the claim to prove (5.7).*

This finishes the proof of the lemma. □

5.2.4 *Marking algorithms*

Consider a cache of size k and fix a request sequence σ. We divide the request sequence into *phases* as follows: phase 0 is the empty sequence. For every $i \geq 1$, phase i is the maximal sequence following phase $i-1$ that contains at most k distinct page requests; that is, if it exists, phase $i+1$ begins on the request that constitutes the $k+1$ distinct page request since the start of the i-th phase. Such a partition is called a *k-phase partition.* This partition is well defined and is independent of any particular algorithm processing σ.

For example, a 3-phase partition:

$$\underbrace{1,2,1,2,1,2,3}_{\text{3-phase \#1}},\underbrace{4,5,6,6,6,6,6,6,6,4,5,4}_{\text{3-phase \#2}},\underbrace{7,7,7,7,1,2}_{\text{3-phase \#3}}.$$

Let σ be any request sequence and consider its k-phase partition. Associate with each page a bit called the *mark.* The marking is done for the sake of analysis (this is not implemented by the algorithm, but "by us" to keep track of the doings of the algorithm). For each page, when its mark bit is set we say that the page is *marked,* and otherwise, *unmarked.*

Suppose that at the beginning of each k-phase we unmark all the pages, and we mark a page when it is first requested during the k-phase. A *marking algorithm* never evicts a marked page from its fast memory.

For example, suppose that $k = 2$, and σ is a request sequence. We show the 2-phases of σ:

$$\sigma = \underbrace{1,1,3,1}_{\text{2-phase \#1}},\underbrace{5,1,5,1,5,1}_{\text{2-phase \#2}},\underbrace{3,4,4,4}_{\text{2-phase \#3}},\underbrace{2,2,2,2}_{\text{2-phase \#4}}. \tag{5.8}$$

See figure 5.6 to examine the marking in this example. Note that after each phase, every page is unmarked and we begin marking afresh, and except for the last phase, all phases are always complete (they have exactly k distinct requests, 2 in this case).

With a marking algorithm, once a request for page p in phase i is made, p stays in the cache until the end of phase i—the first time p is requested, it is marked, and it stays marked for the entire phase, and a marking algorithm never evicts a marked page.

The intuition is that marking algorithms are good schemes for page replacement because, in any given phase, there are at most k distinct pages, so they all fit in a cache of size k, so it does not make sense to evict them in that phase (we can only lose by evicting, as the evicted page might be requested again).

step	1	2	3	4	5	step	1	2	3	4	5
1	x					10	x				x
2	x					11			x		
3	x		x			12			x	x	
4	x		x			13			x	x	
5					x	14			x	x	
6	x				x	15		x			
7	x				x	16		x			
8	x				x	17		x			
9	x				x	18		x			

Fig. 5.6 Marking in example (5.8).

Theorem 5.20. LRU *is a marking algorithm*

Proof. We argue by contradiction; suppose that LRU on a cache of size k is not a marking algorithm. Let σ be a request sequence where there exists a k-phase partition, during which some marked page p is evicted. Consider the first request for p during this k-phase:

$$\sigma = p_1, p_2, p_3, \ldots, \ldots, \underbrace{\ldots, p, \ldots, \ldots, \ldots}_{k\text{-phase}}, \ldots, \ldots$$

Immediately after p is serviced, it is marked as the most recently used page in the cache (i.e., it is put at the top of the doubly-linked list).

In order for p to leave the cache, LRU must incur a page fault while p is the least recently used page. It follows that during the k-phase in question, $k+1$ distinct pages were requested: there are the $k-1$ pages that pushed p to the end of the list, there is p, and the page that got p evicted. Contradiction; a k-phase has at most k distinct pages. □

5.2.5 *FWF*

Flush When Full (FWF) is a very naïve page replacement algorithm that works as follows: whenever there is a page fault and there is no space left in the cache, evict all pages currently in the cache—call this action a "flush."

More precisely, we consider the following version of the FWF algorithm: each slot in the cache has a single bit associated with it. At the beginning, all these bits are set to zero. When a page p is requested, FWF checks only the slots with a marked bit. If p is found, it is serviced. If p is not found, then it has to be brought in from the slow memory (even if it actually is in the cache, in an unmarked slot). FWF looks for a slot with a zero bit, and one of the following happens: (1) a slot with a zero bit (an unmarked page) is found, in which case FWF replaces that page with p. (2) a slot with a zero bit is not found (all pages are marked), in which case FWF unmarks all the slots, and replaces any page with p, and it marks p's bit.

Problem 5.21. *Show that* FWF *is a marking algorithm. Show that* FIFO *is not a marking algorithm.*

Problem 5.22. *A page replacement algorithm* ALG *is* conservative *if, on any consecutive input subsequence containing k or fewer distinct page requests,* ALG *will incur k or fewer page faults. Prove that* LRU *and* FIFO *are conservative, but* FWF *is not.*

5.2.6 *LFD*

The optimal page replacement algorithm turns out to be LFD (*Longest Forward Distance*—see figure 5.2). LFD evicts the page that will not be used for the longest period of time, and as such, it cannot be implemented in practice because it requires knowledge of the future. However, it is very useful for comparison studies, i.e., competitive analysis.

Theorem 5.23. LFD *is the optimal (offline) page replacement algorithm, i.e.,* OPT = LFD.

Proof. We will show that if ALG is any paging algorithm (online or offline), then on any sequence of requests σ, ALG(σ) $\geq$ LFD(σ). As usual, ALG(σ) denotes the number of page faults of ALG on the sequence of requests σ. We assume throughout that all algorithms are working with a cache of a fixed size k. We need to prove the following claim.

Claim 5.24. *Let* ALG *be any paging algorithm. Let $\sigma = p_1, p_2, \ldots, p_n$ be any request sequence. Then, it is possible to construct an offline algorithm* ALG$_i$ *that satisfies the following three properties:*

(1) ALG$_i$ *processes the first $i-1$ requests of σ exactly as* ALG *does,*
(2) if the i-th request results in a page fault, ALG$_i$ *evicts from the cache the page with the "longest forward distance,"*
(3) ALG$_i$(σ) $\leq$ ALG(σ)

Proof. Divide σ into three segments as follows:

$$\sigma = \sigma_1, p_i, \sigma_2,$$

where σ_1 and σ_2 each denote a block of requests.

Recall the proof of theorem 5.11 where we simulated ALG with ALG′ by running a "ghost simulation" of the contents of the cache of ALG on a set of registers, so ALG′ would know what to do with its cache based on the contents of those registers. We do the same thing here: ALG_i runs a simulation of ALG on a set of registers.

As on σ_1, ALG_i is just ALG, it follows that $\text{ALG}_i(\sigma_1) = \text{ALG}(\sigma_1)$, and also, they both do or do not incur a page fault on p_i. If they do not, then let ALG_i continue behaving just like ALG on σ_2, so that $\text{ALG}_i(\sigma) = \text{ALG}(\sigma)$.

However, if they do incur a page fault on p_i, ALG_i evicts the page with the longest forward distance from its cache, and replaces it with p_i. If ALG also evicts the same page, then again, let ALG_i behave just like ALG for the rest of σ, so that $\text{ALG}_i(\sigma) = \text{ALG}(\sigma)$.

Finally, suppose that they both incur a fault at p_i, but ALG evicts some page q and ALG_i evicts some page p, and $p \neq q$; see figure 5.7. If both $p, q \notin \sigma_2$, then let ALG_i behave just like ALG, except the slots with p and q are interchanged (that is, when ALG evicts from the q-slot, ALG_i evicts from the p-slot, and when ALG evicts from the p-slot, ALG_i evicts from the q-slot).

If $q \in \sigma_2$ but $p \notin \sigma_2$, then again let ALG_i, when forced with an eviction, act just like ALG with the two slots interchanged. Note that in this case it may happen that $\text{ALG}_i(\sigma_2) < \text{ALG}(\sigma_2)$, since ALG evicted q, which is going to be requested again, but ALG_i evicted p which will never be requested.

Problem 5.25. *Why the case $q \notin \sigma_2$ and $p \in \sigma_2$ is* not *possible?*

Otherwise, we can assume that ALG_i evicts page p and ALG evicts page q, $p \neq q$, and:

$$\sigma_2 = p_{i+1}, \ldots, q, \ldots, p, \ldots, p_n. \tag{5.9}$$

ALG:		$\not{q}$			p	
ALG_i:		q			$\not{p}$	

Fig. 5.7 ALG evicts q and ALG_i evicts p, denoted with $\not{q}$ and $\not{p}$, respectively, and they both replace their evicted page with p_i.

Assume that the q shown in (5.9) is the earliest instance of q in σ_2. As before, let ALG_i act just like ALG with the q-slot and p-slot interchanged. We know for sure that ALG will have a fault at q. Suppose ALG does not have a fault at p; then, ALG never evicted p, so ALG_i never evicted q, so ALG_i did not have a fault at q. Therefore, $\text{ALG}_i(\sigma_2) \leq \text{ALG}(\sigma_2)$. □

We now show how to use claim 5.24 to prove that LFD is in fact the optimal algorithm. Let $\sigma = p_1, p_2, \ldots, p_n$ be any sequence of requests. By the claim, we know that: $\text{ALG}_1(\sigma) \leq \text{ALG}(\sigma)$. Applying the claim again, we get $(\text{ALG}_1)_2(\sigma) \leq \text{ALG}_1(\sigma)$. Define $\overline{\text{ALG}}_j$ to be $(\cdots((\text{ALG}_1)_2)\cdots)_j$. Then, we obtain that $\overline{\text{ALG}}_j(\sigma) \leq \overline{\text{ALG}}_{j-1}(\sigma)$.

Note that $\overline{\text{ALG}}_n$ acts just like LFD on σ, and therefore we have that $\text{LFD}(\sigma) = \overline{\text{ALG}}_n(\sigma) \leq \text{ALG}(\sigma)$, and we are done. □

Henceforth, OPT can be taken to be synonymous with LFD.

Theorem 5.26. *Any marking algorithm* ALG *is* $\left(\frac{k}{k-h+1}\right)$*-competitive, where k is the size of its cache, and h is the size of the cache of* OPT.

Proof. Fix any request sequence σ and consider its k-phase partition. Assume, for now, that the last phase of σ is complete (in general, the last phase may be incomplete).

Claim 5.27. *For any phase $i \geq 1$, a marking algorithm* ALG *incurs at most k page faults.*

Proof. This follows because there are k distinct page references in each phase. Once a page is requested, it is marked and therefore cannot be evicted until the phase has been completed. Consequently, ALG cannot fault twice on the same page. □

If we denote the i-th k-phase of σ by σ_i, we can express the above claim as $\text{ALG}(\sigma_i) \leq k$. Thus, if there are s phases, $\text{ALG}(\sigma) \leq s \cdot k$.

Claim 5.28. $\text{OPT}(\sigma) \geq s \cdot (k - h + 1)$, *where again we assume that the requests are $\sigma = \sigma_1, \sigma_2, \ldots, \sigma_s$, where σ_s is complete.*

Proof. Let p_a be the first request of phase i, and p_b the last request of phase i. Suppose first that phase $i+1$ exists (that is, i is *not* the last phase). Then, we partition σ into k-phases (even though the cache of OPT is of size k, we still partition σ into k-phases):

$$\sigma = \ldots, p_{a-1}, \underbrace{p_a, p_{a+1}, \ldots, p_b}_{k\text{-phase \#}i}, \underbrace{p_{b+1}, \ldots, \ldots}_{k\text{-phase \#}i+1}$$

After processing request p_a, OPT has at most $h-1$ pages in its cache, not including p_a. From (and including) p_{a+1} until (and including) p_{b+1}, there are at least k distinct requests. Therefore, OPT must incur at least $k-(h-1)=k-h+1$ faults on this segment. To see this, note that there are two cases.

Case 1. p_a appears again in $p_{a+1},\ldots,p_{b+1}$; then there are at least $(k+1)$ distinct requests in the segment $p_{a+1},\ldots,p_{b+1}$, and since OPT has a cache of size h, regardless of the contents of the cache, there will be at least $(k+1)-h=k-h+1$ page faults.

Case 2. Suppose that p_a does *not* appear again in $p_{a+1},\ldots,p_{b+1}$, then since p_a is requested at the beginning of phase i, it is for sure in the cache by the time we start servicing $p_{a+1},\ldots,p_{b+1}$. Since it is not requested again, it is taking up a spot in the cache, so at most $(h-1)$ slots in the cache can be taken up by some of the elements requested in $p_{a+1},\ldots,p_{b+1}$; so again, we have at least $k-(h-1)=k-h+1$ many faults.

If i is the last phase (so $i=s$), we do not have p_{b+1}, so we can only say that we have at least $k-h$ faults, but we make it up with p_1 which has not been counted. □

It follows from claims 5.27 and 5.28 that:

$$\text{ALG}(\sigma)\leq s\cdot k \quad \text{and} \quad \text{OPT}(\sigma)\geq s\cdot(k-h+1),$$

so that:

$$\frac{\text{ALG}(\sigma)}{s\cdot k}\leq 1\leq\frac{\text{OPT}(\sigma)}{s\cdot(k-h+1)},$$

so finally:

$$\text{ALG}(\sigma)\leq\left(\frac{k}{k-h+1}\right)\cdot\text{OPT}(\sigma).$$

In the case that σ can be divided into s complete phases.

As was mentioned above, in general, the last phase may not be complete. Then, we repeat this analysis with $\sigma=\sigma_1,\sigma_2,\ldots,\sigma_{s-1}$, and for σ_s we use α at the end, so we get:

$$\text{ALG}(\sigma)\leq\left(\frac{k}{k-h+1}\right)\cdot\text{OPT}(\sigma)+\alpha$$

Problem 5.29. *Work this out.*

Therefore, in either case we obtain that any marking algorithm ALG is $\left(\frac{k}{k-h+1}\right)$-competitive. □

Problem 5.30. *Implement in Python all the disciplines in table 5.2. Judge them experimentally, by running them on a string of random requests, and plotting their costs—compared to* LFD.

5.3 Answers to selected problems

Problem 5.1. Think of the filing cabinet mentioned at the beginning of this chapter. As we scan the filing cabinet while searching for a particular file, we keep a pointer at a given location along the way (i.e., we "place a finger" as a bookmark in that location) and then insert the accessed file in that location almost free of additional search or reorganization costs. We also assume that it would not make sense to move the file to a later location. Finally, any permutation can be written out as a product of transpositions (check any abstract algebra textbook).

Problem 5.3. The answer is 3. Note that in a list of n items there are $\binom{n}{2} = \frac{n \cdot (n-1)}{2}$ *un*ordered pairs (and $n \cdot (n-1)$ ordered pairs), so to compute Φ, we enumerate all those pairs, and increase a counter by 1 (starting from 0) each time we encounter an inversion. For the second question, note that while we do not know how OPT works exactly, we know that it services σ with the optimal cost, i.e., it services σ in the cheapest way possible. Thus, we can find OPT(σ) by an exhaustive enumeration: given our list $x_1, x_2, \ldots, x_l$ and a sequence of requests $\sigma = \sigma_1, \sigma_2, \ldots, \sigma_n$, we build a tree where the root is labeled with $x_1, x_2, \ldots, x_l$, and the children of the root are all the $l!$ permutations of the list. Then each node in turn has $l!$ many children; the depth of the tree is n. We calculate the cost of each branch and label the leaves with those costs. The cost of each branch is the sum of the costs of all the transpositions required to produce each consecutive node, and the costs of the searches associated with the corresponding list configurations. The cheapest branch (and there may be several) is precisely OPT(σ).

Problem 5.5. The number of elements before x in MTF is $(k-1)$, since x is in the k-th position. Of these $(k-1)$ elements, v are $*$. Both lists contain exactly the same elements, and the non-$*$ before x in MTF, $(k-1-v)$ must all be before x in OPT (if an element is before x in MTF and after x in OPT, then by definition it would be a $*$).

Problem 5.6. In the case of a paid transposition, the only change in the number of inversions can come from the two transposed items, as the relative order with respect to all the other items remains the same. In the case of a free transposition, we know that MTF already put the transposed item x at the front of its list, and we know that free transpositions can only move x forward, so the number of items before x in OPT decreases by 1.

Problem 5.8. OPT is the optimal offline algorithm, and hence it must do at least as well as any algorithm ALG. Suppose we service all requests one-by-one in the naïve way, without making any rearrangements. The cost of this scheme is bounded about by $n \cdot l$, the number of requests times the length of the list. Hence, $\text{OPT}(\sigma) \leq n \cdot l$.

Problem 5.9. By theorem 5.2 we know that

$$\text{MTF}(\sigma) \leq 2 \cdot \text{OPT}_S(\sigma) + \text{OPT}_P(\sigma) - \text{OPT}_F(\sigma) - n$$

and the RHS is

$$\leq 2 \cdot \text{OPT}_S(\sigma) + \text{OPT}_P(\sigma) \leq 2 \cdot (\text{OPT}_S(\sigma) + \text{OPT}_P(\sigma)) = 2 \cdot \text{OPT}(\sigma).$$

This shows that MTF is 2-competitive (with $\alpha = 0$). For the second part, we repeat the above argument, but without "losing" the n factor, so we have $\text{MTF}(\sigma) \leq 2 \cdot \text{OPT}(\sigma) - n$. On the other hand, $\text{OPT}(\sigma) \leq n \cdot l$ (by problem 5.8), so

$$2 \cdot \text{OPT}(\sigma) - n \leq \left(2 - \frac{1}{l}\right) \cdot \text{OPT}(\sigma)$$

Problem 5.12. In the proof of theorem 5.11 we define a matching between the page moves (from slow memory into the cache) of ALG and ALG$'$. In order to show that the matching is one-to-one we postulate the existence of a pair i, j, $i \neq j$, with the following properties: (i) there exists a page p such that $\sigma_i = \sigma_j = p$, (ii) ALG$'$ incurs a page fault at σ_i and σ_j, and (iii) ALG$'$ has to move p into the cache to service σ_i and σ_j and those two moves are matched with the same move of p by ALG. For the sake of the argument in the proof of theorem 5.11 we want the "smallest" such pair—so we use the Least Number Principle (see page 3) to show that if such pairs exist at all, there must exist pairs where $i + j$ is minimal; we take any such pair.

Problem 5.13. Consider the following list:

$$\underbrace{1, 2, 3 \ldots, i, i+1}_{1}, \underbrace{1, 2, 3, \ldots, i-1}_{2}, \underbrace{i+2}_{3}, \underbrace{1, 2, 3, \ldots, i, i+1, i+2}_{4}.$$

If we have a cache of size $i + 1$, then we incur $i + 1$ faults in segment 1 (because the cache is initially empty), then we have $i - 1$ hits in segment 2, then we have another page fault in segment 3 so we evict 1, and in segment 4 we lag behind by 1 all the way, so we incur $i + 2$ page faults. Hence, we incur $i + 1 + 1 + i + 2 = 2i + 4$ page faults in total.

Suppose now that we have a cache of size i. Then we incur $i+1$ page faults in segment 1, then we have $i-1$ page faults in segment 2, and one page fault in segment 3, hence $2i+1$ page faults before starting segment 4. When segment 4 starts, we already have pages 1 through $i-1$ in the cache, so we have hits, and then when $i+1$ is requested, we have a fault, and when $i+2$ is requested we have a hit, and hence only one fault in segment 4. Therefore, we have $2i+2$ page faults with a cache of size i. To understand this solution, make sure that you keep track of the contents of the cache after each of the four segments has been processed. Note that i has to be at least 3 for this example to work.

Problem 5.15. The problem with a singly-linked list is that to find the predecessor of a page we need to start the search always at the beginning of the list, increasing the overhead of maintaining the stack.

Problem 5.21. FWF really implements the marking bit, so it is almost a marking algorithm by definition. FIFO is not a marking algorithm because with $k = 3$, and the request sequence $1, 2, 3, 4, 2, 1$ it will evict 2 in the second phase even though it is marked.

Problem 5.22. We must assume that the cache is of size k. Otherwise the claim is not true: for example, suppose that we have a cache of size 1, and the following sequences: $1, 2, 1, 2$. Then, in that sequence of 4 requests there are only 2 distinct requests, yet with a cache of size 1, there would be 4 faults, for *any* page-replacement algorithm. With a cache of size k, LRU is never going to evict a page during this consecutive subsequence, once the page has been requested. Thus, each distinct page request can only cause one fault. Same goes for FIFO. Thus, they are both conservative algorithms. However, it is possible that half-way through the consecutive subsequence, the cache of FWF is going to get full, and FWF is going to evict everybody. Hence, FWF may have more than one page fault on the same page during this consecutive subsequence.

Problem 5.25. If $p \in \sigma_2$ and $q \notin \sigma_2$, then q would have a "longer forward distance" than p, and so p would not have been evicted by ALG. Rather, ALG would have evicted q or some other page that was not to be requested again.

5.4 Notes

A very complete text book on online algorithms is [Borodin and El-Yaniv (1998)]. See also [Dorrigiv and López-Ortiz (2009)] from the SIGACT news online algorithms column.

Chapter 6

Randomized Algorithms

It is very interesting that we can design procedures which, when confronted with a profusion of choices, instead of laboriously examining all the possible answers to those choices, they flip a coin to decide which way to go, and still "tend to" obtain the right output.

Obviously we save time when we resort to randomness, but what is very surprising is that the output of such procedures can be meaningful. That is, there are problems that computationally appear very difficult to solve, but when allowed the use of randomness it is possible to design procedures that solve those hard problems in a satisfactory manner: the output of the procedure is correct with a small probability of error. In fact this error can be made so small that it becomes negligible (say 1 in 2^{100}—the estimated number of atoms in the observable universe). Thus, many experts believe that the definition of "feasibly computable" ought to be computable in polynomial time with randomness, rather than just in polynomial time.

The advent of randomized algorithms came with the problem of primality testing, which in turn was spurred by the then burgeoning field of cryptography. Historically the first such algorithm was due to [Solovay and Strassen (1977)]. Primality testing remains one of the best problems to showcase the power of randomized algorithms; in this chapter we present the Rabin-Miller algorithm that came after the Solovay-Strassen algorithm, but it is somewhat simpler. Interestingly enough, there is now known a polynomial time algorithm for primality testing (due to [Agrawal *et al.* (2004)]), but the Rabin-Miller algorithm is still used in practice as it is more efficient (and its probability of error is negligible).

In this chapter we present three examples of randomized algorithms: an algorithm for perfect matching, for string pattern matching and finally the Rabin-Miller algorithm. We close with a discussion of cryptography.

6.1 Perfect matching

Consider a bipartite graph $G = (V \cup V', E)$, where $E \subseteq V \times V'$, and its adjacency matrix is defined as follows: $(A_G)_{ij} = x_{ij}$ if $(i, j') \in E_G$, and $(A_G)_{ij} = 0$ otherwise. See the example given in figure 6.1.

Let S_n be the set of all the permutations of n elements. More precisely, S_n is the set of bijections $\sigma : [n] \longrightarrow [n]$. Clearly, $|S_n| = n!$, and it is a well known result from algebra that any permutation $\sigma \in S_n$ can be written as a product of transpositions, that is, permutations that simply exchange two elements in $[n]$, and leave every other element fixed. Any permutation in S_n may be written as a product of transpositions, and although there are many ways to do this (i.e., a representation by transpositions is not unique), the parity of the number of transpositions is constant for any given permutation σ. Let $\text{sgn}(\sigma)$ be 1 or -1, depending on whether the parity of σ is even or odd, respectively.

Recall the Lagrange formula for the determinant:

$$\det(A) = \sum_{\sigma \in S_n} \text{sgn}(\sigma) \prod_{i=1}^{n} A_{i\sigma(i)}. \tag{6.1}$$

Problem 6.1. *Let $G = (V \cup V', E)$ be a graph where $n = |V| = |V'|$ and $E \subseteq V \times V'$. Then, the graph G has a* perfect matching *(i.e., each vertex in V can be paired with a unique vertex in V') iff it is the case that* $\det(A_G) = \sum_{\sigma \in S_n} \text{sgn}(\sigma) \prod_{i=1}^{n} (A_G)_{i\sigma(i)} \neq 0$.

Since $|S_n| = n!$, computing the summation over all the σ in S_n, as

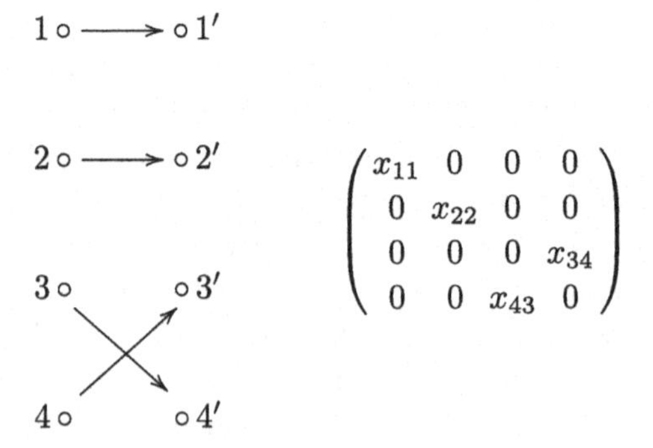

$$\begin{pmatrix} x_{11} & 0 & 0 & 0 \\ 0 & x_{22} & 0 & 0 \\ 0 & 0 & 0 & x_{34} \\ 0 & 0 & x_{43} & 0 \end{pmatrix}$$

Fig. 6.1 On the left we have a bipartite graph $G = (V \cup V', E)$ where $V = \{1, 2, 3, 4\}$, $V' = \{1', 2', 3', 4'\}$ and $E \subseteq V \times V'$, $E = \{(1, 1'), (2, 2'), (3, 4'), (4, 3')\}$. On the right we have the corresponding adjacency matrix A_G.

in (6.1), is computationally very expensive, so we randomly assign values to the x_{ij}'s. The integer determinant, unlike the symbolic determinant, can be computed very efficiently—for example with Berkowitz's algorithm. Let $A_G(x_1, \ldots, x_m)$, $m = |E_G|$, be A_G with its variables renamed to $x_1, \ldots, x_m$. Note that $m \leq n^2$ and each x_l represents some x_{ij}. We obtain a randomized algorithm for the perfect matching problem—see algorithm 6.1.

Algorithm 6.1 Perfect matching

```
Choose m random integers i_1, ..., i_m in {1, ..., M} where M = 2m
compute the integer determinant of A_G(i_1, ..., i_m)
if det(A_G(i_1, ..., i_m)) ≠ 0 then
        return yes, G has a perfect matching
else
        return no, G probably has no perfect matching
end if
```

Algorithm 6.1 is a polynomial time *Monte Carlo algorithm*: "yes" answers are reliable and final, while "no" answers are in danger of a *false negative*. The false negative can arise as follows: G may have a perfect matching, but $(i_1, \ldots, i_m)$ may happen to be a root of the polynomial $\det(A_G(x_1, \ldots, x_m))$. However, the probability of a false negative (i.e., the probability of error) can be made negligibly small, as we shall see shortly.

In line 1 of algorithm 6.1 we say, somewhat enigmatically, "choose m random numbers." How do we "choose" these random numbers? It turns out that the answer to this question is not easy, and obtaining a source of randomness is the Achilles heel of randomized algorithms. We have the science of *pseudo-random number generators* at our disposal, and other approaches, but this formidable topic lies outside the scope of this book, and so we shall naïvely assume that we have a "some source of randomness."

We want to show the correctness of our randomized algorithm, so we need to show that the probability of error is negligible. We start with the Schwarz-Zipper lemma.

Lemma 6.2 (Schwarz-Zippel). *Consider polynomials over* $\mathbb{Z}$, *and let* $p(x_1, \ldots, x_m) \neq 0$ *be a polynomial, where the degree of each variable is* $\leq d$ *(when the polynomial is written out as a sum of monomials), and let* $M > 0$. *Then the number of* m*-tuples* $(i_1, \ldots, i_m) \in \{1, 2, \ldots, M\}^m$ *such that* $p(i_1, \ldots, i_m) = 0$ *is* $\leq mdM^{m-1}$.

Proof. Induction on m (the number of variables). If $m = 1$, $p(x_1)$ can have at most $d = 1 \cdot d \cdot M^{1-1}$ many roots, by the Fundamental Theorem of Algebra.

Suppose the lemma holds for $(m-1)$, and now we want to give an upper bound of mdM^{m-1} on the number of tuples $(i_1, \ldots, i_m)$ such that $p(i_1, \ldots, i_m) = 0$. First we write $p(x_1, \ldots, x_m)$ as $y_d x_m^d + \cdots + y_0 x_m^0$, where each coefficient $y_i = y_i(x_1, \ldots, x_{m-1}) \in \mathbb{Z}[x_1, \ldots, x_{m-1}]$.

So how many tuples $(i_1, \ldots, i_m)$ such that $p(i_1, \ldots, i_m) = 0$ are there? We partition such tuples into two sets: those that set $y_d = 0$ and those that do not. The result is bounded above by the sum of the upper bounds of the two sets; we now give those upper bounds.

Set 1. By the induction hypothesis, y_d is zero for at most $(m-1)dM^{m-2}$ many $(i_1, \ldots, i_{m-1})$ tuples, and x_m can take M values, and so $p(x_1, \ldots, x_m)$ is zero for at most $(m-1)dM^{m-1}$ tuples. Note that we are over-counting here; we are taking *all* tuples that set $y_d = 0$.

Set 2. For each combination of M^{m-1} values for $x_1, \ldots, x_{m-1}$, there are at most d roots of the resulting polynomial (again by the Fundamental Theorem of Algebra), i.e., dM^{m-1}. Note that again we are over-counting as some of those settings to the $x_1, \ldots, x_m$ will result in $y_d = 0$.

Adding the two estimates gives us mdM^{m-1}. □

Lemma 6.3. *Algorithm 6.1 is correct.*

Proof. We want to show that algorithm 6.1 for perfect matching is a reliable Monte Carlo algorithm, which means that "yes" answers are 100% correct, while "no" answers admit a negligible probability of error.

If the algorithm answers "yes," then $\det(A_G(i_1, \ldots, i_m)) \neq 0$ for some randomly selected $i_1, \ldots, i_m$, but then the symbolic determinant $\det(A_G(x_1, \ldots, x_m)) \neq 0$, and so, by problem 6.1, G has a perfect matching. So "yes" answers indicate with absolute certainty that there is a perfect matching.

Suppose that the answer is "no." Then we apply lemma 6.2 to $\det(A_G(x_1, \ldots, x_m))$, with $M = 2m$, and obtain that the probability of a false negative is

$$\leq \frac{m \cdot d \cdot M^{m-1}}{M^m} = \frac{m \cdot 1 \cdot (2m)^{m-1}}{(2m)^m} = \frac{m}{2m} = \frac{1}{2}.$$

Now suppose we perform "many independent experiments," meaning that we run algorithm 6.1 k many times, each times choosing a random set $i_1, \ldots, i_m$. Then, if the answer *always* comes zero we know that the probability of error is $\leq \left(\frac{1}{2}\right)^k = \frac{1}{2^k}$. For $k = 100$, the error becomes *negligible.* □

In the last paragraph of the proof of lemma 6.3 we say that we run algorithm 6.1 k many times, and so bring down the probability of error to being less than $\frac{1}{2^k}$, which for $k = 100$ is truly negligible. Running the algorithm k times to get the answer is called *amplification* (because we decrease drastically the probability of error, and so amplify the certainty of having a correct answer); note that the beauty of this approach is that while we run the algorithm only k times, the probability of error goes down exponentially quickly to $\frac{1}{2^k}$. Just to put things in perspective, if $k = 100$, then $\frac{1}{2^{100}}$ is so minuscule that by comparison the probability of earth being hit by a large meteor—while running the algorithm—is a virtual certainty (and being hit by a large meteor would spear anyone the necessity to run algorithms in the first place).

Problem 6.4. *Show how to use algorithm 6.1 to actually find a perfect matching.*

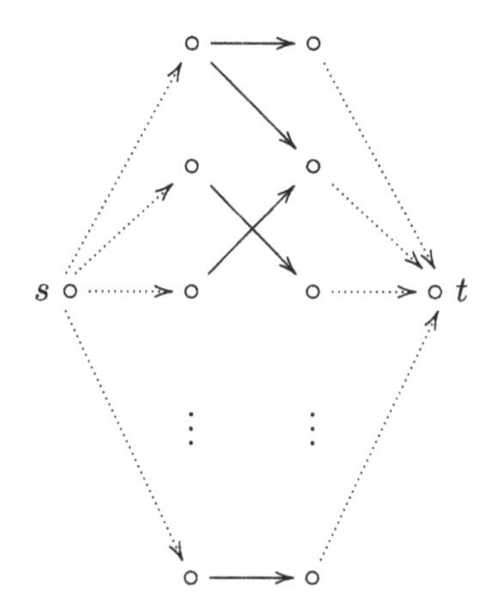

Fig. 6.2 Reduction of perfect matching to max flow.

Perfect matching can be easily reduced[1] to a "max flow problem": as an example, consider the perfect matching problem given in figure 6.1; add two new nodes s, t, and connect s to all the nodes in the left-column of the matching problem, and connect t to all the nodes in the right-column of the matching problem, and give each edge a capacity of 1, and ask if there is a flow $\geq n$ (where n is the number of nodes in each of the two components

[1]Recall that we have examined briefly the idea of *reductions* on page 89.

of the given bipartite graph) from s to t; see figure 6.2.

As the max flow problem can be solved in polynomial time without using randomness, it follows that perfect matching can also be solved in polynomial time without randomness. Still, the point of this section was to exhibit a simple randomized algorithm, and that we have accomplished.

6.2 Pattern matching

In this section we design a randomized algorithm for pattern matching. Consider the set of strings over $\{0,1\}$, and let $M : \{0,1\} \longrightarrow M_{2\times 2}(\mathbb{Z})$, that is, M is a map from strings to 2×2 matrices over the integers ($\mathbb{Z}$) defined as follows:

$$M(\varepsilon) = \begin{bmatrix} 1 & 0 \\ 0 & 1 \end{bmatrix}; \quad M(0) = \begin{bmatrix} 1 & 0 \\ 1 & 1 \end{bmatrix}; \quad M(1) = \begin{bmatrix} 1 & 1 \\ 0 & 1 \end{bmatrix},$$

and for strings $x, y \in \{0,1\}^*$, $M(xy) = M(x)M(y)$, where the operation on the left-hand side is concatenation of strings, and the operation on the right-hand side is multiplication of matrices.

Problem 6.5. *Show that $M(x)$ is well defined, that is, no matter how we evaluate M on x we always get the same result. Also show that M is one-to-one.*

Problem 6.6. *Show that for $x \in \{0,1\}^n$, the entries of $M(x)$ are bounded by the n-th Fibonacci number. Recall that Fibonacci numbers were defined in problem 1.5.*

By considering the matrices $M(x)$ modulo a suitable prime p, i.e., by taking all the entries of $M(x)$ modulo a prime p, we perform efficient randomized pattern matching. We wish to determine whether x is a substring of y, where $|x| = n$, $|y| = m$, $n \leq m$. Define

$$y(i) = y_i y_{i+1} \cdots y_{n+i-1},$$

for appropriate $i \in \{1, \ldots, m-n+1\}$. Select a prime $p \in \{1, \ldots, nm^2\}$, and let $A = M(x) \pmod p$ and $A(i) = M(y(i)) \pmod p$. Note that

$$A(i+1) = M^{-1}(y_i)A(i)M(y_{n+i}) \pmod p,$$

which makes the computation of subsequent $A(i)$'s efficient.

What is the probability of getting a false positive? It is the probability that $A(i) = M(y(i)) \pmod p$ even though $A(i) \neq M(y(i))$. This is less

Algorithm 6.2 Pattern matching

Pre-condition: $x, y \in \{0,1\}^*$, $|x| = n, |y| = m$ and $n \leq m$

1: select a random prime $p \leq nm^2$
2: $A \longleftarrow M(x) \pmod{p}$
3: $B \longleftarrow M(y(1)) \pmod{p}$
4: **for** $i = 1, \ldots, m - n + 1$ **do**
5:     **if** $A = B$ **then**
6:         **if** $x = y(i)$ **then**
7:             **return** found a match at position i
8:         **end if**
9:     **end if**
10:     $B \longleftarrow M^{-1}(y_i) \cdot B \cdot M(y_{n+i})$
11: **end for**

than the probability that $p \in \{1, \ldots, nm^2\}$ divides a (non-zero) entry in $A(i) - M(y(i))$. Since these entries are bounded by $F_n < 2^n$, less than n distinct primes can divide any of them. On the other hand, there are $\pi(nm^2) \approx (nm^2)/(\log(nm^2))$ primes in $\{1, \ldots, nm^2\}$ (by the Prime Number Theorem). So the probability of a false positive is $O(1/m)$.

Note that algorithm 6.2 has no error; it is randomized, but all potential answers are checked for a *false positive* (in line 6). Checking for these potential candidates is called *fingerprinting.* The idea of fingerprinting is to check only those substrings that "look" like good candidates, making sure that when we "sniff" for a candidate we never miss the solution (in this case, if $x = y(i)$, for some i, then $y(i)$ will always be a candidate). On the other hand, there may be j's such that $x \neq y(j)$ and yet they are candidates; but the probability of that is small. The use of randomness in algorithm 6.2 just lowers the average time complexity of the procedure; such algorithms are called *Las Vegas algorithms.*

6.3 Primality testing

One way to determine whether a number p is prime, is to try all possible numbers $n < p$, and check if any are divisors[2]. Obviously, this brute force procedure has exponential time complexity in the length of p, and so it has a prohibitive time cost. Although a polytime (deterministic) algorithm

[2]This section requires a little bit of number theory; see appendix A for all the necessary background.

for primality is now known (see [Agrawal *et al.* (2004)]), the Rabin-Miller randomized algorithm for primality testing is simpler and more efficient, and therefore still used in practice.

Fermat's Little theorem (see theorem A.6 in appendix A) provides a "test" of sorts for primality, called the Fermat test; the Rabin-Miller algorithm (algorithm 6.3) is based on this test. When we say that p passes the Fermat test at a, what we mean is that $a^{(p-1)} \equiv 1 \pmod{p}$. Thus, all primes pass the Fermat test for all $a \in \mathbb{Z}_p - \{0\}$.

Unfortunately, there are also composite numbers n that pass the Fermat tests for every $a \in \mathbb{Z}_n^*$; these are the so called *Carmichael numbers*, for example, 561, 1105, 1729, etc.

Lemma 6.7. *If p is a composite non-Carmichael number, then it passes at most half of the tests in $\mathbb{Z}_p^*$. That is, if p is a composite non-Carmichael number, then for at most half of the a's in the set $\mathbb{Z}_p^*$ it is the case that $a^{(p-1)} \equiv 1 \pmod{p}$.*

Proof. We say that a is a *witness* for p if a fails the Fermat test for p, that is, a is a witness if $a^{(p-1)} \not\equiv 1 \pmod{p}$. Let $S \subseteq \mathbb{Z}_p^*$ consist of those elements $a \in \mathbb{Z}_p^*$ for which $a^{p-1} \equiv 1 \pmod{p}$. It is easy to check that S is in fact a subgroup of $\mathbb{Z}_p^*$. Therefore, by Lagrange theorem (theorem A.10), $|S|$ must divide $|\mathbb{Z}_p^*|$. Suppose now that there exists an element $a \in \mathbb{Z}_p^*$ for which $a^{p-1} \not\equiv 1 \pmod{p}$. Then, $S \neq \mathbb{Z}_p^*$, so the next best thing it can be is "half" of $\mathbb{Z}_p^*$, so $|S|$ must be at most half of $|\mathbb{Z}_p^*|$. □

Problem 6.8. *Give an alternative proof of lemma 6.7 without using group theory.*

A number is *pseudoprime* if it is either prime or Carmichael. The last lemma suggests an algorithm for pseudoprimeness: on input p, check whether $a^{(p-1)} \equiv 1 \pmod{p}$ for some random $a \in \mathbb{Z}_p - \{0\}$. If p fails this test (i.e., $a^{(p-1)} \not\equiv 1 \pmod{p}$), then p is composite for sure. If p passes the test, then p is probably pseudoprime. We show that the probability of error in this case is $\leq \frac{1}{2}$. Suppose p is not pseudoprime. If $\gcd(a,p) \neq 1$, then $a^{(p-1)} \not\equiv 1 \pmod{p}$ (by proposition A.4), so assuming that p passed the test, it must be the case that $\gcd(a,p) = 1$, and so $a \in \mathbb{Z}_p^*$. But then, by lemma 6.7, at least half of the elements of $\mathbb{Z}_p^*$ are witnesses of non-pseudoprimeness.

Problem 6.9. *Show that if* $\gcd(a,p) \neq 1$ *then* $a^{(p-1)} \not\equiv 1 \pmod{p}$.

The informal algorithm for pseudoprimeness described in the paragraph above is the basis for the Rabin-Miller algorithm which we discuss next. The Rabin-Miller algorithm extends the pseudoprimeness test to deal with Carmichael numbers.

Algorithm 6.3 Rabin-Miller

```
1:  If n = 2, accept; if n is even and n > 2, reject.
2:  Choose at random a positive a in Z_n.
3:  if a^(n-1) ≢ 1 (mod n) then
4:          reject
5:  else
6:          Find s, h such that s is odd and n − 1 = s2^h
7:          Compute the sequence a^(s·2^0), a^(s·2^1), a^(s·2^2), ..., a^(s·2^h) (mod n)
8:          if all elements in the sequence are 1 then
9:                  accept
10:         else if the last element different from 1 is −1 then
11:                 accept
12:         else
13:                 reject
14:         end if
15: end if
```

Note that this is a polytime (randomized) algorithm: computing powers $\pmod{n}$ can be done efficiently with *repeated squaring*—for example, if $(n-1)_b = c_r \dots c_1 c_0$, then compute

$$a_0 = a, a_1 = a_0^2, a_2 = a_1^2, \dots, a_r = a_{r-1}^2 \pmod{n},$$

and so $a^{n-1} = a_0^{c_0} a_1^{c_1} \cdots a_r^{c_r} \pmod{n}$. Thus obtaining the powers in lines 6 and 7 is not a problem.

Problem 6.10. *Implement the Rabin-Miller algorithm in Python. In the first naïve version, the algorithm should run on integer inputs (the built in* `int` *type). In the second, more sophisticated version, the algorithm should run on inputs which are numbers encoded as binary strings, with the trick of repeated squaring in order to cope with large numbers.*

Theorem 6.11. *If* n *is a prime then the Rabin-Miller algorithm accepts it; if* n *is composite, then the algorithm rejects it with probability* $\geq \frac{1}{2}$.

Proof. If n is prime, then by Fermat's Little theorem $a^{(n-1)} \equiv 1 \pmod{n}$, so line 4 cannot reject n. Suppose that line 13 rejects n; then there exists a b in $\mathbb{Z}_n$ such that $b \not\equiv \pm 1 \pmod{n}$ and $b^2 \equiv 1 \pmod{n}$. Therefore, $b^2 - 1 \equiv 0 \pmod{n}$, and hence

$$(b-1)(b+1) \equiv 0 \pmod{n}.$$

Since $b \not\equiv \pm 1 \pmod{n}$, both $(b-1)$ and $(b+1)$ are strictly between 0 and n, and so a prime n cannot divide their product. This gives a contradiction, and therefore no such b exists, and so line 13 cannot reject n.

If n is an odd composite number, then we say that a is a *witness* (of compositness) for n if the algorithm rejects on a. We show that if n is an odd composite number, then at least half of the a's in $\mathbb{Z}_n$ are witnesses. The distribution of those witnesses in $\mathbb{Z}_n$ appears to be very irregular, but if we choose our a at random, we hit a witness with probability $\geq \frac{1}{2}$.

Because n is composite, either n is the power of an odd prime, or n is the product of two odd co-prime numbers. This yields two cases.

Case 1. Suppose that $n = q^e$ where q is an odd prime and $e > 1$. Set $t := 1 + q^{e-1}$. From the binomial expansion of t^n we obtain:

$$t^n = (1 + q^{e-1})^n = 1 + nq^{e-1} + \sum_{l=2}^{n} \binom{n}{l} (q^{e-1})^l, \tag{6.2}$$

and therefore $t^n \equiv 1 \pmod{n}$. If $t^{n-1} \equiv 1 \pmod{n}$, then $t^n \equiv t \pmod{n}$, which from the observation about t and t^n is not possible, hence t is a line 4 witness. But the set of line 4 non-witnesses, $S_1 := \{a \in \mathbb{Z}_n | a^{(n-1)} \equiv 1 \pmod{n}\}$, is a subgroup of $\mathbb{Z}_n^*$, and since it is not equal to $\mathbb{Z}_n^*$ (t is not in it), by Lagrange's theorem S_1 is at most half of $\mathbb{Z}_n^*$, and so it is at most half of $\mathbb{Z}_n$.

Case 2. Suppose that $n = qr$, where q, r are co-prime. Among all line 13 non-witnesses, find a non-witness for which the -1 appears in the largest position in the sequence in line 7 of the algorithm (note that -1 is a line 13 non-witness, so the set of these non-witnesses is not empty). Let x be such a non-witness and let j be the position of -1 in its sequence, where the positions are numbered starting at 0; $x^{s \cdot 2^j} \equiv -1 \pmod{n}$ and $x^{s \cdot 2^{j+1}} \equiv 1 \pmod{n}$. The line 13 non-witnesses are a subset of $S_2 := \{a \in \mathbb{Z}_n^* | a^{s \cdot 2^j} \equiv \pm 1 \pmod{n}\}$, and S_2 is a subgroup of $\mathbb{Z}_n^*$.

By the CRT there exists $t \in \mathbb{Z}_n$ such that

$$\begin{array}{ll} t \equiv x \pmod{q} & \\ t \equiv 1 \pmod{r} \end{array} \Rightarrow \begin{array}{ll} t^{s \cdot 2^j} \equiv -1 \pmod{q} \\ t^{s \cdot 2^j} \equiv 1 \pmod{r} \end{array}$$

Hence t is a witness because $t^{s \cdot 2^j} \not\equiv \pm 1 \pmod{n}$ but on the other hand $t^{s \cdot 2^{j+1}} \equiv 1 \pmod{n}$.

Problem 6.12. *Show that* $t^{s \cdot 2^j} \not\equiv \pm 1 \pmod{n}$.

Therefore, just as in case 1, we have constructed a $t \in \mathbb{Z}_n^*$ which is not in S_2, and so S_2 can be at most half of $\mathbb{Z}_n^*$, and so at least half of the elements in $\mathbb{Z}_n$ are witnesses. □

Problem 6.13. *First show that the sets S_1 and S_2 (in the proof of theorem 6.11) are indeed subgroups of $\mathbb{Z}_n^*$, and that in case 2 all non-witnesses are contained in S_2. Then show that at least half of the elements of $\mathbb{Z}_n$ are witnesses when n is composite, without using group theory.*

Note that by running the algorithm k times on independently chosen a, we can make sure that it rejects a composite with probability $\geq 1 - \frac{1}{2^k}$ (it will always accept a prime with probability 1). Thus, for $k = 100$ the probability of error, i.e., of a false positive, is negligible.

6.4 Public key cryptography

Cryptography has well known applications to security; for example, we can use our credit cards when purchasing online because, when we send our credit card numbers, they are encrypted, and even though they travel through a public channel, no one, but the intended recipient, can read them. Cryptography has also a fascinating history: from the first uses recorded by Herodotus during the Persian wars five centuries BC, to the exploits at Bletchley Park during WWII—the reader interested in the history of cryptography should read the fascinating book [Singh (1999)].

A *Public Key Cryptosystem (PKC)* consists of three sets: K, the set of (pairs of) *keys*, M, the set of *plaintext* messages, and C, the set of *ciphertext* messages. A pair of keys in K is $k = (k_{\text{priv}}, k_{\text{pub}})$; the *private* (or *secret*) key and the *public* key, respectively. For each k_{pub} there is a corresponding *encryption* function $e_{k_{\text{pub}}} : M \longrightarrow C$ and for each k_{priv} there is a corresponding *decryption* function $d_{k_{\text{priv}}} : C \longrightarrow M$.

The property that the encryption and decryption functions must satisfy is that if $k = (k_{\text{priv}}, k_{\text{pub}}) \in K$, then $d_{k_{\text{priv}}}(e_{k_{\text{pub}}}(m)) = m$ for all $m \in M$. The necessary assumption is that it must be difficult to compute $d_{k_{\text{priv}}}(c)$ just from knowing k_{pub} and c. But, with the additional *trapdoor* information k_{priv} it becomes easy to compute $d_{k_{\text{priv}}}(c)$.

In the following sections we present three different encryption schemes; Diffie-Hellman, which is not really a PKC but rather a way of agreeing on a secret key over an insecure channel, as well as ElGamal and RSA. All three require large primes (in practice about 1,000 bit long); a single prime for Diffie-Hellman and ElGamal, and a pair of primes for RSA. How to go about it? The answer will of course involve the Rabin-Miller algorithm from the previous section.

Here is how we go about it: we know by the Prime Number Theorem that there are about $\pi(n) = n/\log n$ many primes $\leq n$. This means that there are $2^n/n$ primes among n-bit integers, roughly 1 in n, and these primes are fairly uniformly distributed. So we pick an integer at random, in a given range, and apply the Rabin-Miller algorithm to it.

6.4.1 *Diffie-Hellman key exchange*

If p is prime, then one can show—though the proof is difficult and we omit it here—that there exists a $g \in \mathbb{Z}_p^*$ such that $\langle g \rangle = \{g^1, g^2, \ldots, g^{p-1}\} = \mathbb{Z}_p^*$. This g is called a *primitive root* for $\mathbb{Z}_p^*$. Given an $h \in \mathbb{Z}_p^*$, the *Discrete Log Problem (DLP)* is the problem of finding an $x \in \{1, \ldots, p-1\}$ such that $g^x \equiv h \pmod{p}$. That is, $x = \log_g(h)$.

For example, $p = 56609$ is a prime number and $g = 2$ is a generator for $\mathbb{Z}_{56609}^*$, that is $\mathbb{Z}_{56609}^* = \{2^1, 2^2, 2^3, \ldots, 2^{56608}\}$, and $\log_2(38679) = 11235$.

Problem 6.14. *If $p = 7$, explain why $g = 3$ would work as a generator for $\mathbb{Z}_p^*$. Is every number in $\mathbb{Z}_7^*$ a generator for $\mathbb{Z}_7^*$?*

The DLP is assumed to be a difficult problem. We are going to use it to set up a way for Alice and Bob to agree on a secret key over an insecure channel. First Alice and Bob agree on a large prime p and an integer $g \in \mathbb{Z}_p^*$. In fact, g does not have to be a primitive root for p; it is sufficient, and much easier, to pick a number g of order roughly $p/2$. See, for example, exercise 1.31 in [Hoffstein *et al.* (2008)]. The numbers p, g are public knowledge, that is, $k_{\text{pub}} = \langle p, g \rangle$.

Then Alice picks a secret a and Bob picks a secret b. Alice computes $A := g^a \pmod{p}$ and Bob computes $B := g^b \pmod{p}$. Then Alice and Bob exchange A and B over an insecure link. On her end, Alice computes $A' := B^a \pmod{p}$ and Bob, on his end, computes $B' := A^b \pmod{p}$. Clearly,

$$A' \equiv_p B^a \equiv_p (g^b)^a \equiv_p g^{ab} \equiv_p (g^a)^b \equiv_p A^b \equiv_p B'.$$

This common value $A' = B'$ is their secret key. Thus, Diffie-Hellman is not really a fully-fledged PKC; it is just a way for two parties to agree on a secret value over an insecure channel. Also note that computing A and B involves computing large powers of g modulo the prime p; if this is done naïvely by multiplying g times itself a many times, then this procedure is impractical for large a. We use repeated squaring instead; see page 131 where we discuss this procedure.

Problem 6.15. *Suppose that Alice and Bob agree on $p = 23$ and $g = 5$, and that Alice's secret is $a = 8$ and Bob's secret is $b = 15$. Show how the Diffie-Hellman exchange works in this case, and what is the secret key that they end up having.*

Suppose that Eve is eavesdropping on this exchange. She is capable of gleaning the following information from it: $\langle p, g, g^a \pmod{p}, g^b \pmod{p}\rangle$. Computing $g^{ab} \pmod{p}$ (i.e., $A' = B'$) from this information is known as the *Diffie-Hellman Problem* (DHP), and it is assumed to be difficult when p is a large prime number.

But suppose that Eve has an efficient way of solving the DLP. Then, from $g^a \pmod{p}$ she computes a, and from $g^b \pmod{p}$ she computes b, and now she can easily compute $g^{ab} \pmod{p}$. On the other hand, it is not known if solving DHP efficiently yields an efficient solution for the DLP.

Problem 6.16. *Consider Shank's algorithm—algorithm 6.4. Show that Shank's algorithm computes x, such that $g^x \equiv_p h$, in time $O(n \log n)$ that is, in time $O(\sqrt{p}\log(\sqrt{p}))$.*

Algorithm 6.4 Shank's babystep-giantstep

Pre-condition: p prime, $\langle g\rangle = \mathbb{Z}_p^*$, $h \in \mathbb{Z}_p^*$

1: $n \longleftarrow 1 + \lfloor\sqrt{p}\rfloor$
2: $L_1 \longleftarrow \{g^0, g^1, g^2, \ldots, g^n\} \pmod{p}$
3: $L_2 \longleftarrow \{hg^0, hg^{-n}, hg^{-2n}, \ldots, hg^{-n^2}\} \pmod{p}$
4: Find $g^i \equiv_p hg^{-jn} \in L_1 \cap L_2$
5: $x \longleftarrow jn + i$
6: **return** x

Post-condition: $g^x \equiv_p h$

Problem 6.17. *Implement algorithm 6.4 in Python.*

While it seems to be difficult to mount a direct attack on Diffie-Hellman, that is, to attack it by solving the related discrete logarithm problem, there is a rather insidious way of breaking it, called "the man-in-the-middle" attack. It consists in Eve taking advantage of the lack of authentication for the parties; that is, how does Bob know that he is receiving a message from Alice, and how does Alice know that she is receiving a message from Bob? Eve can take advantage of that, and intercept a message A from Alice intended for Bob and replace it with $E = g^e \pmod{p}$, and intercept the message B from Bob intended for Alice and also replace it with $E = g^e \pmod{p}$, and from then on read all the correspondence by pretending to be Bob to Alice, and Alice to Bob, translating message encoded with $g^{ae} \pmod{p}$ to message encoded with $g^{be} \pmod{p}$, and vice versa.

Problem 6.18. *Suppose that $f : \mathbb{N} \times \mathbb{N} \longrightarrow \mathbb{N}$ is a function with the following properties:*

- *for all $a, b, g \in \mathbb{N}$, $f(g, ab) = f(f(g, a), b) = f(f(g, b), a)$,*
- *for any g, $h_g(c) = f(g, c)$ is a* one-way function, *that is, a function that is easy to compute, but whose inverse is difficult to compute*[3].

Explain how f could be used for public key crypto in the style of Diffie-Hellman.

6.4.2 *ElGamal*

This is a true PKC, where Alice and Bob agree on public p, g, such that p is a prime and $\mathbb{Z}_p^* = \langle g \rangle$. Alice also has a private a and publishes a public $A := g^a \pmod{p}$. Bob wants to send a message m to Alice, so he creates an *ephemeral* key b, and sends the pair c_1, c_2 to Alice where:

$$c_1 := g^b \pmod{p}; \qquad c_2 := mA^b \pmod{p}.$$

Then, in order to read the message, Alice computes:

$$c_1^{-a} c_2 \equiv_p g^{-ab} m g^{ab} \equiv_p m.$$

Note that to compute c_1^{-a} Alice first computes the inverse of c_1 in $\mathbb{Z}_p^*$, which she can do efficiently using the extended Euclid's algorithm (see algorithm 1.15 or algorithm 3.5), and then computes the a-th power of the result.

More precisely, here is how we compute the inverse of a k in $\mathbb{Z}_n^*$. Observe that if $k \in \mathbb{Z}_n^*$, then $\gcd(k, n) = 1$, so using algorithm 1.15 we obtain s, t such that $sk + tn = 1$, and further s, t can be chosen so that s is in $\mathbb{Z}_n^*$

[3]The existence of such functions is one of the underlying assumptions of cryptography; the discrete logarithm is an example of such a function, but there is no proof of existence, only a well-founded supposition.

To see that, first obtain any s, t, and then just add to s the appropriate number of positive or negative multiples of n to place it in the set $\mathbb{Z}_n^*$, and adjust t by the same number of multiples of opposite sign.

Problem 6.19. *Let $p = 7$ and $g = 3$.*

(1) Let $a = 4$ be Alice's secret key, so

$$A = g^a \pmod{p} = 3^4 \pmod{7} = 4.$$

Let $p = 7, g = 3, A = 4$ be public values.
Suppose that Bob wants to send the message $m = 2$ to Alice, with ephemeral key $b = 5$. What is the corresponding pair $\langle c_1, c_2 \rangle$ that he sends to Alice? Show what are the actual values and how are they computed.

(2) What does Alice do in order to read the message $\langle 5, 4 \rangle$? That is, how does Alice extract m out of $\langle c_1, c_2 \rangle = \langle 5, 4 \rangle$?

Problem 6.20. *We say that we can* break *ElGamal, if we have an efficient way for computing m from $\langle p, g, A, c_1, c_2 \rangle$. Show that we can break ElGamal if and only if we can solve the DHP efficiently.*

Problem 6.21. *Write a Python application which implements the ElGamal digital signature scheme. Your command-line program ought to be invoked as follows:* `sign 11 6 3 7` *and then accept a single line of* `ASCII` *text until the new-line character appears (i.e., until you press enter). That is, once you type* `sign 11 6 3 7` *at the command line, and press return, you type a message: '*`A message`*' and after you have pressed return again, the digital signature, which is going to be a pair of positive integers, will appear below.*

We now explain how to obtain this digital signature: first convert the characters in the string '`A message`*' into the corresponding* `ASCII` *codes, and then obtain a* hash *of those codes by multiplying them all modulo 11; the result should be the single number 5.*

To see this observe the table:

```
A    65   10
     32   1
m    109  10
e    101  9
s    115  1
s    115  5
a    97   1
```

```
g        103    4
e        101    8
.         46    5
```

The first column contains the characters, the second the corresponding `ASCII` *codes, and the i-th entry in the third column contains the product of the first i codes modulo 11. The last entry in the third column is the hash value 5.*

We sign the hash value, i.e., if the message is $m =$ `A message.`, *then we sign hash*$(m) = 5$. *Note that we invoke* `sign` *with four arguments, i.e., we invoke it with* p, g, x, k *(in our running example,* $11, 6, 3, 7$ *respectively).*

Here p *must be a prime,* $1 < g, x, k < p-1$, *and* $\gcd(k, p-1) = 1$. *This is a condition of the input; you don't have to test in your program whether the condition is met—we may assume that it is.*

Now the algorithm signs $h(m)$ *as follows: it computes*

$$r = g^k \pmod p$$
$$s = k^{-1}(h(m) - xr) \pmod{(p-1)}$$

If s *is zero, start over again, by selecting a different* k *(meeting the required conditions). The signature of* m *is precisely the pair of numbers* (r, s). *In our running example we have the following values:*

$$m = \texttt{A message}; \quad h(m) = 5; \quad p = 11; \quad g = 6; \quad x = 3; \quad k = 7$$

and so the signature of '`A message`*' with the given parameters will be:*

$$\begin{aligned} r &= 6^7 \pmod{11} = 8 \\ s &= 7^{-1}(5 - 3 \cdot 8) \pmod{(11-1)} \\ &= 3 \cdot (-19) \pmod{10} \\ &= 3 \cdot 1 \pmod{10} \\ &= 3 \end{aligned}$$

i.e., the signature of '`A message`*' would be* $(r, s) = (8, 3)$.

Problem 6.22. *In problem 6.21:*

(1) *Can you identify the (possible) weaknesses of this digital signature scheme? Can you compose a different message* m' *such that* $h(m) = h(m')$*?*
(2) *If you receive a message* m, *and a signature pair* (r, s), *and you only know* p, g *and* $y = g^x \pmod p$, *i.e.,* p, g, y *are the* public *information, how can you "verify" the signature—and what does it mean to verify the signature?*

(3) *Research on the web a better suggestion for a hash function h.*
(4) *Show that when used without a (*good*) hash function, ElGamal's signature scheme is* existentially forgeable*; i.e., an adversary Eve can construct a message m and a valid signature (r, s) for m.*
(5) *In practice k is a random number; show that it is absolutely necessary to choose a new random number for each message.*
(6) *Show that in the verification of the signature it is essential to check whether $1 \leq r \leq p-1$, because otherwise Eve would be able to sign message of her choice, provided she knows one valid signature (r, s) for some message m, where m is such that $1 \leq m \leq p-1$ and $\gcd(m, p-1) = 1$.*

6.4.3 *RSA*

Choose two odd primes p, q, and set $n = pq$. Choose $k \in \mathbb{Z}^*_{\phi(n)}$, $k > 1$. Advertise f, where $f(m) \equiv m^k \pmod{n}$. Compute l, the inverse of k in $\mathbb{Z}^*_{\phi(n)}$. Now $\langle n, k \rangle$ are public, and the key l is secret, and so is the function g, where $g(C) \equiv C^l \pmod{n}$. Note that $g(f(m)) \equiv_n m^{kl} \equiv_n m$.

Problem 6.23. *Show that $m^{kl} \equiv m \pmod{n}$. In fact there is an implicit assumption about m in order for this to hold; what is this assumption?*

Problem 6.24. *Observe that we could break RSA if factoring were easy.*

We now make two observations about the security of RSA. The first one is that the primes p, q cannot be chosen "close" to each other. To see what we mean, note that:

$$n = \left(\frac{p+q}{2}\right)^2 - \left(\frac{p-q}{2}\right)^2.$$

Since p, q are close, we know that $s := (p-q)/2$ is small, and $t := (p+q)/2$ is only slightly larger than $\sqrt{n}$, and $t^2 - n = s^2$ is a perfect square. So we try the following candidate values for t:

$$\lceil\sqrt{n}\rceil + 0, \quad \lceil\sqrt{n}\rceil + 1, \quad \lceil\sqrt{n}\rceil + 2, \ldots$$

until $t^2 - n$ is a perfect square s^2. Clearly, if s is small, we will quickly find such a t, and then $p = t + s$ and $q = t - s$.

The second observation is that if were to break RSA by computing l efficiently from n and k, then we would be able to factor n in randomized polynomial time. Since $\phi(n) = \phi(pq) = (p-1)(q-1)$, it follows that:

$$\begin{aligned} p+q &= n-\phi(n)+1 \\ pq &= n, \end{aligned} \tag{6.3}$$

and from these two equations we obtain:

$$(x-p)(x-q) = x^2-(p+q)x+pq = x^2-(n-\phi(n)+1)x+n.$$

Thus, we can compute p, q by computing the roots of this last polynomial. Using the classical quadratic formula $x = (-b \pm \sqrt{b^2-4ac})/2a$, we obtain that p, q are:

$$\frac{(n-\phi(n)+1) \pm \sqrt{(n-\phi(n)+1)^2-4n}}{2}.$$

Suppose that Eve is able to compute l from n and k. If Eve knows l, then she knows that whatever $\phi(n)$ is, it divides $kl-1$, and so she has equations (6.3) but with $\phi(n)$ replaced with $(kl-1)/a$, for some a. This a can be computed in randomized polynomial time, but we do not present the method here. Thus, the claim follows.

If Eve is able to factor she can obviously break RSA; on the other hand, if Eve can break RSA—by computing l from n, k—, then she would be able to factor in randomized polytime.

On the other hand, Eve may be able to break RSA *without* computing l, so the preceding observation do not imply that breaking RSA is as hard as factoring.

6.5 Further exercises

There is a certain reversal of priorities in cryptography, in that difficult problem become allies, rather than obstacles. On page 81 we mentioned NP-hard problems, which are problems for which there are no feasible solutions when the instances are "big enough."

The Simple Knapsack Problem (SKS) (see section 4.3) is one such problem, and we can use it to define a cryptosystem. The *Merkle-Hellman subset-sum cryptosystem* is based on SKS, and it works as follows. First, Alice creates a secret key consisting of the following elements:

- A *super-increasing sequence*: $\mathbf{r} = (r_1, r_2, \ldots, r_n)$ where $r_i \in \mathbb{N}$, and the property of being "super-increasing" refers to $2r_i \leq r_{i+1}$, for all $1 \leq i < n$.

- A pair of positive integers A, B with two conditions: $2r_n < B$ and $\gcd(A, B) = 1$.

The public key consists of $\mathbf{M} = (M_1, M_2, \ldots, M_n)$ where $M_i = Ar_i \pmod{B}$.

Suppose that Bob wants to send a plain-text message $x \in \{0,1\}^n$, i.e., x is a binary string of length n. Then he uses Alice's public key to computer $S = \sum_{i=1}^{n} x_i M_i$, where x_i is the i-th bit of x, interpreted as integer 0 or 1. Bob now sends S to Alice.

For Alice to read the message she computes $S' = A^{-1}S \pmod{B}$, and she solves the subset-sum problem S' using the super-increasing $\mathbf{r}$. The subset-sum problem, for a general sequence $\mathbf{r}$, is very difficult, but when $\mathbf{r}$ is super-increasing (note that $\mathbf{M}$ is assumed not to be super-increasing!) the problem can be solved with a simple greedy algorithm.

More precisely, Alice finds a subset of $\mathbf{r}$ whose sum is precisely S'. Any subset of $\mathbf{r}$ can be identified with a binary string of length n, by assuming that x_i is 1 iff r_i is in this subset. Hence Alice "extracts" x out of S'.

For example, let $\mathbf{r} = (3, 11, 24, 50, 115)$, and $A = 113$, $B = 250$. Check that all conditions are met, and verify that $\mathbf{M} = (89, 243, 212, 150, 245)$. To send the secret message $x = 10101$, we compute

$$S = 1 \cdot 89 + 0 \cdot 243 + 1 \cdot 212 + 0 \cdot 150 + 1 \cdot 245 = 546.$$

Upon receiving S, we multiply it times 177, the inverse of 113 in mod 250, and obtain 142. Now x may be extracted out of 142 with a simple greedy algorithm.

Problem 6.25.

(1) *Show that if $\mathbf{r} = (r_1, r_2, \ldots, r_n)$ is a super-increasing sequence then $r_{i+1} > \sum_{j=1}^{i} r_j$, for all $1 \le i < n$.*
(2) *Suppose that $\mathbf{r} = (r_1, r_2, \ldots, r_n)$ is a super-increasing sequence, and suppose that there is a subset of $\mathbf{r}$ whose sum is S. Provide a (natural) greedy algorithm for computing this subset, and show that your algorithm is correct.*

Problem 6.26. *Implement the Merkle-Hellman subset-sum cryptosystem in Python. Call the program* `sscrypt`, *and it should work with three switches:* `-e -d -v`, *for* encrypt, decrypt *and* verify. *That is,*

`sscrypt -e` $M_1\ M_2\ \ldots\ M_n\ x$

encrypts the string $x = x_1x_2\ldots x_n \in \{0,1\}^n$ with the public key given by $\mathbf{M} = (M_1, M_2, \ldots, M_n)$, and outputs S. On the other hand, `sscrypt -d` $r_1\ r_2\ \ldots\ r_n\ A\ B\ S$

decrypts the string $x = x_1x_2\ldots x_n \in \{0,1\}^n$ *from* S *using the secret key given by* $\mathbf{r} = (r_1, r_2, \ldots, r_n)$ *and* A, B*; that is, it outputs* x *on input* $\mathbf{r}, A, B, S$*. Finally,*

`sscrypt -v` $r_1\ r_2\ \ldots\ r_n\ A\ B$

checks that $\mathbf{r} = (r_1, r_2, \ldots, r_n)$ *is super-increasing, it checks that* $2r_n < B$ *and that* $\gcd(A, B) = 1$*, and outputs the corresponding public key given by* $\mathbf{M} = (M_1, M_2, \ldots, M_n)$.

6.6 Answers to selected problems

Problem 6.4. We use algorithm 6.1 to find perfect matching (if one exists) as follows: pick $1 \in V$, and consider each $(1, i') \in E$ in turn, remove it from G to obtain $G_{1,i'} = ((V - \{1\}) \cup (V' - \{i'\}), E_{1,i'})$, where $E_{1,i'}$ consists of all the edges of E except those adjacent on 1 or i', until for some $i' \in V'$ we obtain a $G_{1,i'}$ for which the algorithm answers "yes." Then we know that there is a perfect matching that matches 1 and i'. Continue with $G_{1,i'}$.
Problem 6.5. $M(x)$ is well defined because matrix multiplication is associative. We now show that $M(x) = M(y)$ implies that $x = y$ (i.e., the map M is one-to-one). Given $M = M(x)$ we can "decode" x uniquely as follows: if the first column of M is greater than the second (where the comparison is made component-wise), then the last bit of x is zero, and otherwise it is 1. Let M' be M where we subtract the smaller column from the larger, and repeat.
Problem 6.6. For a given string x, $M(x_1x_2\ldots x_n)$ is such that the "smaller" column is bounded by f_{n-1} and the "larger" column is bounded by f_n. We can show this inductively: the basis case, $x = x_1$, is obvious. For the inductive step, assume it holds for $x \in \{0,1\}^n$, and show it still holds for $x \in \{0,1\}^{n+1}$: this is clear as whether x_{n+1} is 0 or 1, one column is added to the other, and the other column remains unchanged.
Problem 6.9. Suppose that $\gcd(a,p) \neq 1$. By proposition A.4 we know that if $\gcd(a,p) \neq 1$, then a does not have a (multiplicative) inverse in $\mathbb{Z}_p$. Thus, it is not possible for $a^{(p-1)} \equiv 1 \pmod p$ to be true, since then it would follow that $a \cdot a^{(p-2)} \equiv 1 \pmod p$, and hence a would have a (multiplicative) inverse.
Problem 6.12. To see why $t^{s\cdot 2^j} \not\equiv \pm 1 \pmod n$ observe the following: suppose that $a \equiv -1 \pmod q$ and $a \equiv 1 \pmod r$, where $\gcd(q,r) = 1$. Suppose that $n = qr|(a+1)$, then $q|(a+1)$ and $r|(a+1)$, and since $r|(a-1)$

as well, it follows that $r|[(a+1)-(a-1)]$, so $r|2$, so $r=2$, so n must be even, which is not possible since we deal with even n's in line 1 of the algorithm.

Problem 6.13. Showing that S_1, S_2 are subgroups of $\mathbb{Z}_n^*$ is easy; it is obvious in both cases that 1 is there, and closure and existence of inverse can be readily checked.

To give the same proof without group theory, we follow the cases in the proof of theorem 6.11. Let t be the witness constructed in case 1. If d is a (stage 3) non-witness, we have $d^{p-1} \equiv 1 \pmod{p}$, but then $dt \pmod{p}$ is a witness. Moreover, if d_1, d_2 are distinct (stage 3) non-witnesses, then $d_1 t \not\equiv d_2 t \pmod{p}$. Otherwise, $d_1 \equiv_p d_1 \cdot t \cdot t^{p-1} \equiv_p d_2 \cdot t \cdot t^{p-1} \equiv_p d_2$. Thus the number of (stage 3) witnesses must be at least as large as the number of non-witnesses.

We do the same for case 2; let d be a non-witness. First, $d^{s \cdot 2^j} \equiv \pm 1 \pmod{p}$ and $d^{s \cdot 2^{j+1}} \equiv 1 \pmod{p}$ owing to the way that j was chosen. Therefore $dt \pmod{p}$ is a witness because $(dt)^{s \cdot 2^j} \not\equiv \pm 1 \pmod{p}$ and $(dt)^{s \cdot 2^{j+1}} \equiv 1 \pmod{p}$.

Second, if d_1 and d_2 are distinct non-witnesses, $d_1 t \not\equiv d_2 t \pmod{p}$. The reason is that $t^{s \cdot 2^{j+1}} \equiv 1 \pmod{p}$. Hence $t \cdot t^{s \cdot 2^{j+1}-1} \equiv 1 \pmod{p}$. Therefore, if $d_1 t \equiv d_2 t \pmod{p}$, then $d_1 \equiv_p d_1 t \cdot t^{s \cdot 2^{j+1}-1} \equiv_p d_2 t \cdot t^{s \cdot 2^{j+1}-1} \equiv_p d_2$. Thus in case 2, as well, the number of witnesses must be at least as large as the number of non-witnesses.

Problem 6.14. $3^1 = 3, 3^2 = 9 = 2, 3^3 = 2 \cdot 3 = 6, 3^4 = 6 \cdot 3 = 4, 3^5 = 4 \cdot 3 = 5, 3^6 = 5 \cdot 3 = 1$, all computations $\pmod 7$, and thus $g = 3$ generates all numbers in $\mathbb{Z}_7^*$. Not every number is a generator: for example, 4 is not.

Problem 6.15. Alice and Bob agree to use a prime $p = 23$ and base $g = 5$. Alice chooses secret $a = 8$; sends Bob $A = g^a \pmod{p}$

$$A = 5^8 \pmod{23} = 16$$

Bob chooses secret $b = 15$; sends Alice $B = g^b \pmod{p}$

$$B = 5^{15} \pmod{23} = 19$$

Alice computes $s = B^a \pmod{p}$

$$s = 19^8 \pmod{23} = 9$$

Bob computes $s = A^b \pmod{p}$

$$s = 16^{15} \pmod{23} = 9$$

As can be seen, both end up having $s = 9$, their shared secret key.

Problem 6.18. Suppose that we have a one-way function as in the question. First Alice and Bob agree on a public g and exchange it (the eavesdropper knows g therefore). Then, let Alice generate a secret a and let Bob generate a secret b. Alice sends $f(g,a)$ to Bob and Bob sends $f(g,b)$ to Alice. Notice that because h_g is one-way, an eavesdropper cannot get a or b from $h_g(a) = f(g,a)$ and $h_g(b) = f(g,b)$. Finally, Alice computes $f(f(g,b),a)$ and Bob computes $f(f(g,a),b)$, and by the properties of the function both are equal to $f(g,ab)$ which is their secret shared key. The eavesdropper cannot compute $f(g,ab)$ feasibly.

Problem 6.19. For the first part,

$$\begin{aligned} c_1 &= g^b \pmod p = 3^5 \pmod 7 = 5 \\ c_2 &= mA^b \pmod p = 2 \cdot 4^5 \pmod 7 = 2 \cdot 2 = 4 \end{aligned}$$

For the second part,

$$\begin{aligned} m &= c_1^{-a} c_2 \pmod p \\ &= 5^{-4} 4 \pmod 7 \\ &= (5^{-1})^4 4 \pmod 7 \\ &= 3^4 4 \pmod 7 \\ &= 4 \cdot 4 \pmod 7 \\ &= 2 \end{aligned}$$

Problem 6.20. The DHP on input $\langle p, g, A \equiv_p g^a, B \equiv_p g^b \rangle$ outputs $g^{ab} \pmod p$, and the ElGamal problem, call it ELGP, on input

$$\langle p, g, A \equiv_p g^a, c_1 \equiv_p g^b, c_2 \equiv_p mA^b \rangle \tag{6.4}$$

outputs m. We want to show that we can break Diffie-Hellman, i.e., solve DHP efficiently, if and only if we can break ElGamal, i.e., solve ELGP efficiently. The key-word here is *efficiently*, meaning in polynomial time.

($\Rightarrow$) Suppose we can solve DHP efficiently; we give an efficient procedure for solving ELGP: given the input (6.4) to ELGP, we obtain $g^{ab} \pmod p$ from $A \equiv_p g^a$ and $c_1 \equiv g^b$ using the efficient solver for DHP. We then use the extended Euclidean algorithm, see problem 1.21—and note that the extended Euclid's algorithm runs in polynomial time, to obtain $(g^{ab})^{-1} \pmod p$. Now,

$$c_2 \cdot (g^{ab})^{-1} \equiv_p mg^{ab}(g^{ab})^{-1} \equiv_p m = m$$

where the last equality follows from $m \in \mathbb{Z}_p$.

($\Leftarrow$) Suppose we have an efficient solver for the ELGP. To solve the DHP, we construct the following input to ELGP:

$$\langle p, g, A \equiv_p g^a, c_1 \equiv_p g^b, c_2 = 1 \rangle.$$

Note that $c_2 = 1 \equiv_p \underbrace{(g^{ab})^{-1}}_{=m} A^b$, so using the efficient solver for ELGP we obtain $m \equiv_p (g^{ab})^{-1}$, and now using the extended Euclid's algorithm we obtain the inverse of $(g^{ab})^{-1} \pmod{p}$, which is just $g^{ab} \pmod{p}$, so we output that.

Problem 6.22.

(1) The weakness of our scheme lies in the hash function, which computes the same hash values for different messages, and in fact it is easy to find messages with the same hash value—for example, by adding pairs of letters (anywhere in the message) such that their corresponding ASCII values are inverses modulo p.
Examples (from the assignments) of messages with the same hash value are: "A mess" and "L message." In general, by its nature, any hash function is going to have such *collisions*, i.e., messages such that:

$$h(\text{A message.}) = h(\text{A mess}) = h(\text{L message}) = 5,$$

but there are hash functions which are *collision-resistant* in the sense that it is computationally hard to find two messages m, m' such that $h(m) = h(m')$. A good hash function is also a *one-way function* in the sense that given a value y it is computationally hard to find an m such that $h(m) = y$.

(2) Verifying the signature means checking that it was the person in possession of x that signed the document m. Two subtle things: first we say "in possession of x" rather than the "legitimate owner of x," simply because x may have been compromised (for example stolen). Second, and this is why this scheme is so brilliant, we can check that "someone in possession of x" signed the message, even *without knowing what x is*! We know y, where $y = g^x \pmod{p}$, but for large p, it is difficult to compute x from y (this is called the Discrete Log Problem, DLP).
Here is how we verify that "someone in possession of x" signed the message m. We check $0 < r < p$ and $0 < s < p - 1$ (see Q6), and we compute $v := g^{h(m)} \pmod{p}$ and $w := y^r r^s \pmod{p}$; g, p are public, m is known, and the function $h : \mathbb{N} \longrightarrow [p-1]$ is also

known, and r, s is the given signature. If v and w match, then the signature is valid.
To see that this works note that we defined $s := k^{-1}(h(m) - xr) \pmod{p-1}$. Thus, $h(m) = xr + sk \pmod{p-1}$. Now, Fermat's Little Theorem (FLT—see page 114 in the textbook), says that $g^{p-1} = 1 \pmod p$, and therefore

$$g^{h(m)} \stackrel{(*)}{=} g^{xr+sh} = (g^x)^r(g^k)^s = y^r r^s \pmod p.$$

The FLT is applied in the $(*)$ equality: since $h(m) = xr + sk \pmod{p-1}$ it follows that $(p-1)|(h(m) - (xr + sk))$, which means that $(p-1)z = h(m) - (xr + sk)$ for some z, and since $g^{(p-1)z} = (g^{(p-1)})^z = 1^z = 1 \pmod p$, it follows that $g^{h(m)-(xr+sk)} = 1 \pmod p$, and so $g^{h(m)} = g^{xr+sk} \pmod p$.

(3) Here are the hash functions implemented by `GPG`, version 1.4.9: `MD5, SHA1, RIPEMD160, SHA256, SHA384, SHA512, SHA224.`

(4) To see this, let b, c be numbers such that $\gcd(c, p-1) = 1$. Set $r = g^b y^c$, $s = -rc^{-1} \pmod{p-1}$ and $m = -rbc^{-1} \pmod{p-1}$. Then (m, r, s) satisfies $g^m = y^r r^s$. Since in practice a hash function h is applied to the message, and it is the hash value that is really signed, to forge a signature for a meaningful message is not so easy. An adversary has to find a meaningful message $\tilde{m}$ such that $h(\tilde{m}) = h(m)$, and when h is collision-resistant this is hard.

(5) If the same random number k is used in two different messages $m \neq m'$, then it is possible to compute k as follows: $s - s' = (m - m')k^{-1} \pmod{p-1}$, and hence $k = (s - s')^{-1}(m - m') \pmod{p-1}$.

(6) Let m' be a message of Eve's choice, $u = m'm^{-1} \pmod{p-1}$, $s' = su \pmod{p-1}$, r' and integer such that $r' = r \pmod p$ and $r' = ru \pmod{p-1}$. This r' can be obtained by the so called Chinese Reminder Theorem (see theorem A.14). Then (m', r', s') is accepted by the verification procedure.

Problem 6.23. Why $m^{kl} \equiv_n m$? Observe that $kl = 1 + (-t)\phi(n)$, where $(-t) > 0$, and so $m^{kl} \equiv_n m^{1+(-t)\phi(n)} \equiv_n m \cdot (m^{\phi(n)})^{(-t)} \equiv_n m$, because $m^{\phi(n)} \equiv_n 1$. Note that this last statement does not follow directly from Euler's theorem (theorem A.13), because $m \in \mathbb{Z}_n$, and not necessarily in $\mathbb{Z}_n^*$. Note that to make sure that $m \in \mathbb{Z}_n^*$ it is enough to insist that we have $0 < m < \min\{p, q\}$; so we break a large message into small pieces.

It is interesting to note that we can bypass Euler's theorem, and just use Fermat's Little theorem: we know that $m^{(p-1)} \equiv_p 1$ and $m^{(q-1)} \equiv_q 1$, so

$m^{(p-1)(q-1)} \equiv_p 1$ and $m^{(q-1)(p-1)} \equiv_q 1$, thus $m^{\phi(n)} \equiv_p 1$ and $m^{\phi(n)} \equiv_q 1$. This means that $p|(m^{\phi(n)} - 1)$ and $q|(m^{\phi(n)} - 1)$, so, since p, q are distinct primes, it follows that $(pq)|(m^{\phi(n)} - 1)$, and so $m^{\phi(n)} \equiv_n 1$.

Problem 6.24. If factoring integers were easy, RSA would be easily broken: if we were able to factor n, we would obtain the primes p, q, and hence it would be easy to compute $\phi(n) = \phi(pq) = (p-1)(q-1)$, and from this we obtain l, the inverse of k.

Problem 6.25. We show that $\forall i \in [n-1]$ it is the case that $r_{i+1} \sum_{j=1}^{i} r_j$ by induction on i. The basis case is $i - 1$, so

$$r_2 \geq 2r_1 > r_1 = \sum_{j=1}^{i} r_j,$$

where $r_2 \geq 2r_1$ by the property of being super-increasing. For the induction step we have

$$r_{i+1} \geq 2r_i = r_i + r_i > r_i + \sum_{j=1}^{i-1} r_j = \sum_{j=1}^{i} r_j,$$

where we used the property of being super-increasing and the induction hypothesis.

Here is the algorithm for the second question:

$X \longleftarrow S$
$Y \longleftarrow \emptyset$
for $i = n \ldots 1$ **do**
  if $(r_i \leq X)$ **then**
    $X \longleftarrow X - r_i$
    $Y \longleftarrow Y \cup \{i\}$
  end if
end for

---

and let the pre-condition state that $\{r_i\}_{i=1}^{n}$ is super-increasing and that there exists an $\mathcal{S} \subseteq \{r_i\}_{i=1}^{n}$ such that $\sum_{i \in \mathcal{S}} r_i = S$. Let the post-condition state that $\sum_{i \in Y} r_i = S$.

Define the following loop invariant: "Y is promising" in the sense that it can be extended, with indices of weights not considered yet, into a solution. That is, after considering i, there exists a subset E of $\{i-1, \ldots, 1\}$ such that $\sum_{j \in X \cup E} r_j = S$.

The basis case is trivial since initially $X = \emptyset$, and since the pre-condition guarantees the existence of a solution, X can be extended into that solution.

For the induction step, consider two cases. If $r_i > X$ then i is not added, but Y can be extended with $E' \subseteq \{i-1, i-2, \ldots, 1\}$. The reason is that by induction hypothesis X was extended into a solution by some $E \subseteq \{i, i-1, \ldots, 1\}$ and i was not part of the extension as r_i was too big to fit with what was already in Y, i.e., $E' = E$.

If $r_i \leq X$ then $i \in E$ since by previous part the remaining weights would not be able to close the gap between S and $\sum_{j \in Y} r_j$.

6.7 Notes

This chapter is based on chapter 8 in [Soltys (2009)]. For a discussion of Berkowitz's algorithm see [Berkowitz (1984); Soltys (2002)].

Most algorithms books provide a chapter on the *Min-Cut Max-Flow* problem mentioned in the context of perfect matching; see for example chapter 7 in [Kleinberg and Tardos (2006)].

For a discussion on generating random numbers see, for example, chapter 7 in [Press *et al.* (2007)].

Credit for inventing the Monte Carlo method often goes to Stanisław Ulam, a Polish born mathematician who worked for John von Neumann on the United States Manhattan Project during World War II. Ulam is primarily known for designing the hydrogen bomb with Edward Teller in 1951. He invented the Monte Carlo method in 1946 while pondering the probabilities of winning a card game of solitaire.

Section 6.2 is based on [Karp and Rabin (1987)].

It was the randomized test for primality that stirred interest in randomized computation in the late 1970's. Historically, the first randomized algorithm for primality was given by [Solovay and Strassen (1977)]; a good exposition of this algorithm, with all the necessary background, can be found in §11.1 in [Papadimitriou (1994)], and another in §18.5 in [von zur Gathen and Gerhard (1999)].

R. D. Carmichael first noted the existence of the Carmichael numbers in 1910, computed fifteen examples, and conjectured that though they are infrequent there were infinitely many. In 1956, Erdös sketched a technique for constructing large Carmichael numbers ([Hoffman (1998)]), and a proof was given by [Alford *et al.* (1994)] in 1994.

The first three Carmichael numbers are 561, 1105, 1729, where the last number shown on this list is called the Hardy-Ramanujan number, after a famous anecdote of the British mathematician G. H. Hardy regarding a hospital visit to the Indian mathematician Srinivasa Ramanujan. Hardy wrote: *I remember once going to see him when he was ill at Putney. I had ridden in taxi cab number 1729 and remarked that the number seemed to me rather a dull one, and that I hoped it was not an unfavorable omen. "No," he replied, "it is a very interesting number; it is the smallest number expressible as the sum of two cubes in two different ways."*.

Section 6.4 is based on material from [Hoffstein *et al.* (2008)] and [Delfs and Knebl (2007)].

RSA is named after the first letters of the last names of its inventors: Ron **R**ivest, Adi **S**hamir, and Leonard **A**dleman.

Appendix A

Number Theory and Group Theory

In this section we work with the set of integers and natural numbers

$$\mathbb{Z} = \{\dots, -3, -2, -1, 0, 1, 2, 3, \dots\}, \quad \mathbb{N} = \{0, 1, 2, \dots\}.$$

We say that x *divides* y, and write $x|y$ if $y = qx$. If $x|y$ we say that x is *divisor* (also *factor*) of y. Using the terminology from section 1.3, $x|y$ if and only if $y = \text{div}(x, y) \cdot x$. We say that a number p is *prime* if its only divisors are itself and 1.

Claim A.1. *If p is a prime, and $p|a_1 a_2 \dots a_n$, then $p|a_i$ for some i.*

Proof. It is enough to show that if $p|ab$ then $p|a$ or $p|b$. Let $g = \gcd(a, p)$. Then $g|p$, and since p is a prime, there are two cases. Case 1, $g = p$, then since $g|a$, $p|a$. Case 2, $g = 1$, so there exist u, v such that $au + pv = 1$ (see algorithm 1.15), so $abu + pbv = b$. Since $p|ab$, and $p|p$, it follows that $p|(abu + pbv)$, so $p|b$. □

Theorem A.2 (Fundamental Theorem of Arithmetic). *For $a \geq 2$, $a = p_1^{e_1} p_2^{e_2} \cdots p_r^{e_r}$, where p_i are prime numbers, and other than rearranging primes, this factorization is unique.*

Proof. We first show the existence of the factorization, and then its uniqueness. The proof of existence is by complete induction; the basis case is $a = 2$, where 2 is a prime. Consider an integer $a > 2$; if a is prime then it is its own factorization (just as in the basis case). Otherwise, if a is composite, then $a = b \cdot c$, where $1 < b, c < a$; apply the induction hypothesis to b and c.

To show uniqueness suppose that $a = p_1 p_2 \dots p_s = q_1 q_2 \dots q_t$ where we have written out all the primes, that is, instead of writing p^e we write $p \cdot p \cdots p$, e times. Since $p_1|a$, it follows that $p_1|q_1 q_2 \dots q_t$. So $p_1|q_j$ for some

j, by claim A.1, but then $p_1 = q_j$ since they are both primes. Now delete p_1 from the first list and q_j from the second list, and continue. Obviously we cannot end up with a product of primes equal to 1, so the two list must be identical. □

Let $m \geq 1$ be an integer. We say that a and b are *congruent modulo* m, and write $a \equiv b \pmod{m}$ (or sometimes $a \equiv_m b$) if $m|(a-b)$. Another way to say this is that a and b have the same remainder when divided by m; using the terminology of section 1.3, we can say that $a \equiv b \pmod{m}$ if and only if $\text{rem}(m, a) = \text{rem}(m, b)$.

Problem A.3. *Show that if* $a_1 \equiv_m a_2$ *and* $b_1 \equiv_m b_2$, *then* $a_1 \pm b_1 \equiv_m a_2 \pm b_2$ *and* $a_1 \cdot b_1 \equiv_m a_2 \cdot b_2$.

Proposition A.4. *If* $m \geq 1$, *then* $a \cdot b \equiv_m 1$ *for some* b *if and only if* $\gcd(a, m) = 1$.

Proof. ($\Rightarrow$) If there exists a b such that $a \cdot b \equiv_m 1$, then we have $m|(ab-1)$ and so there exists a c such that $ab - 1 = cm$, i.e., $ab - cm = 1$. And since $\gcd(a, m)$ divides both a and m, it also divides $ab - cm$, and so $\gcd(a, m)|1$ and so it must be equal to 1.

($\Leftarrow$) Suppose that $\gcd(a, m) = 1$. By the extended Euclid's algorithm (see algorithm 1.15) there exist u, v such that $au+mv = 1$, so $au-1 = -mv$, so $m|(au - 1)$, so $au \equiv_m 1$. So let $b = u$. □

Let $\mathbb{Z}_m = \{0, 1, 2, \ldots, m-1\}$. We call $\mathbb{Z}_m$ the set of integers modulo m. To add or multiply in the set $\mathbb{Z}_m$, we add and multiply the corresponding integers, and then take the remainder of the division by m as the result. Let $\mathbb{Z}_m^* = \{a \in \mathbb{Z}_m | \gcd(a, m) = 1\}$. By proposition A.4 we know that $\mathbb{Z}_m^*$ is the subset of $\mathbb{Z}_m$ consisting of those elements which have multiplicative inverses in $\mathbb{Z}_m$.

The function $\phi(n)$ is called the *Euler totient function*, and it is the number of elements less than n that are co-prime to n, i.e., $\phi(n) = |\mathbb{Z}_n^*|$.

Problem A.5. *If we are able to factor, we are also able to compute* $\phi(n)$. *Show that if* $n = p_1^{k_1} p_2^{k_2} \cdots p_l^{k_l}$, *then* $\phi(n) = \prod_{i=1}^{l} p_i^{k_i - 1}(p_i - 1)$.

Theorem A.6 (Fermat's Little Theorem). *Let* p *be a prime number and* $\gcd(a, p) = 1$. *Then* $a^{p-1} \equiv 1 \pmod{p}$.

Proof. For any a such that $\gcd(a, p) = 1$ the following products

$$1a, 2a, 3a, \ldots, (p-1)a, \tag{A.1}$$

all taken mod p, are pairwise distinct. To see this suppose that $ja \equiv ka \pmod{p}$. Then $(j-k)a \equiv 0 \pmod{p}$, and so $p|(j-k)a$. But since by assumption $\gcd(a,p)=1$, it follows that $p \nmid a$, and so by claim A.1 it must be the case that $p|(j-k)$. But since $j,k \in \{1,2,\ldots,p-1\}$, it follows that $-(p-2) \leq j-k \leq (p-2)$, so $j-k=0$, i.e., $j=k$.

Thus the numbers in the list (A.1) are just a reordering of the list $\{1,2,\ldots,p-1\}$. Therefore

$$a^{p-1}(p-1)! \equiv_p \prod_{j=1}^{p-1} j \cdot a \equiv_p \prod_{j=1}^{p-1} j \equiv_p (p-1)!. \tag{A.2}$$

Since all the numbers in $\{1,2,\ldots,p-1\}$ have inverses in $\mathbb{Z}_p$, as $\gcd(i,p)=1$ for $1 \leq i \leq p-1$, their product also has an inverse. That is, $(p-1)!$ has an inverse, and so multiplying both sides of (A.2) by $((p-1)!)^{-1}$ we obtain the result. □

Problem A.7. *Give a second proof of Fermat's Little theorem using the binomial expansion, i.e.,* $(x+y)^n = \sum_{j=0}^{n} \binom{n}{j} x^j y^{n-j}$ *applied to* $(a+1)^p$.

We say that $(G,*)$ is a *group* if G is a set and $*$ is an operation, such that if $a,b \in G$, then $a*b \in G$ (this property is called *closure*). Furthermore, the operation $*$ has to satisfy the following three properties:

(1) *identity law:* There exists an $e \in G$ such that $e*a=a*e=a$ for all $a \in G$.
(2) *inverse law:* For every $a \in G$ there exists an element $b \in G$ such that $a*b=b*a=e$. This element b is called an *inverse* and it can be shown that it is unique; hence it is often denoted as a^{-1}.
(3) *associative law:* For all $a,b,c \in G$, we have $a*(b*c)=(a*b)*c$.

If $(G,*)$ also satisfies the *commutative law*, that is, if for all $a,b \in G$, $a*b=b*a$, then it is called a *commutative* or *Abelian group*.

Typical examples of groups are $(\mathbb{Z}_n,+)$ (integers mod n under addition) and $(\mathbb{Z}_n^*,\cdot)$ (integers mod n under multiplication). Note that both these groups are Abelian. These are, of course, the two groups of concern for us; but there are many others: $(\mathbb{Q},+)$ is an infinite group (rationals under addition), $\mathrm{GL}(n,\mathbb{F})$ (which is the group of $n \times n$ invertible matrices over a field $\mathbb{F}$), and S_n (the *symmetric group* over n elements, consisting of permutations of $[n]$ where $*$ is function composition).

Problem A.8. *Show that* $(\mathbb{Z}_n,+)$ *and* $(\mathbb{Z}_n^*,\cdot)$ *are groups, by checking that the corresponding operation satisfies the three axioms of a group.*

We let $|G|$ denote the number of elements in G (note that G may be infinite, but we are concerned mainly with finite groups). If $g \in G$ and $x \in \mathbb{N}$, then $g^x = g * g * \cdots * g$, x times. If it is clear from the context that the operation is $*$, we use juxtaposition ab instead of $a * b$.

Suppose that G is a finite group and $a \in G$; then the smallest $d \in \mathbb{N}$ such that $a^d = e$ is called the *order* of a, and it is denoted as $\text{ord}_G(a)$ (or just $\text{ord}(a)$ if the group G is clear from the context).

Proposition A.9. *If G is a finite group, then for all $a \in G$ there exists a $d \in \mathbb{N}$ such that $a^d = e$. If $d = \text{ord}_G(a)$, and $a^k = e$, then $d|k$.*

Proof. Consider the list $a^1, a^2, a^3, \ldots$. If G is finite there must exist $i < j$ such that $a^i = a^j$. Then, $(a^{-1})^i$ applied to both sides yields $a^{i-j} = e$. Let $d = \text{ord}(a)$ (by the LNP we know that it must exist!). Suppose that $k \geq d$, $a^k = e$. Write $k = dq + r$ where $0 \leq r < d$. Then $e = a^k = a^{dq+r} = (a^d)^q a^r = a^r$. Since $a^d = e$ it follows that $a^r = e$, contradicting the minimality of $d = \text{ord}(a)$, unless $r = 0$. □

If $(G, *)$ is a group we say that H is a *subgroup* of G, and write $H \leq G$, if $H \subseteq G$ and H is closed under $*$. That is, H is a subset of G, and H is itself a group. Note that for any G it is always the case that $\{e\} \leq G$ and $G \leq G$; these two are called the *trivial subgroups* of G. If $H \leq G$ and $g \in G$, then gH is called a *left coset of* G, and it is simply the set $\{gh | h \in H\}$. Note that gH is not necessarily a subgroup of G.

Theorem A.10 (Lagrange). *If G is a finite group and $H \leq G$, then $|H|$ divides $|G|$, i.e., the order of H divides the order of G.*

Proof. If $g_1, g_2 \in G$, then the two cosets g_1H and g_2H are either identical or $g_1H \cap g_2H = \emptyset$. To see this, suppose that $g \in g_1H \cap g_2H$, so $g = g_1h_1 = g_2h_2$. In particular, $g_1 = g_2h_2h_1^{-1}$. Thus, $g_1H = (g_2h_2h_1^{-1})H$, and since it can be easily checked that $(ab)H = a(bH)$ and that $hH = H$ for any $h \in H$, it follows that $g_1H = g_2H$.

Therefore, for a finite $G = \{g_1, g_2, \ldots, g_n\}$, the collection of sets $\{g_1H, g_2H, \ldots, g_nH\}$ is a partition of G into subsets that are either disjoint or identical; from among all subcollections of identical cosets we pick a representative, so that $G = g_{i_1}H \cup g_{i_2}H \cup \cdots \cup g_{i_m}H$, and so $|G| = m|H|$, and we are done. □

Problem A.11. *Let $H \leq G$. Show that if $h \in H$, then $hH = H$, and that in general for any $g \in G$, $|gH| = |H|$. Finally, show that $(ab)H = a(bH)$.*

Problem A.12. *If G is a group, and $\{g_1, g_2, \ldots, g_k\} \subseteq G$, then the set $\langle g_1, g_2, \ldots, g_k \rangle$ is defined as follows*

$$\{x_1 x_2 \cdots x_p | p \in \mathbb{N}, x_i \in \{g_1, g_2, \ldots, g_k, g_1^{-1}, g_2^{-1}, \ldots, g_k^{-1}\}\}.$$

Show that $\langle g_1, g_2, \ldots, g_k \rangle \leq G$, and it is called the subgroup generated *by $\{g_1, g_2, \ldots, g_k\}$. Also show that when G is finite $|\langle g \rangle| = \mathrm{ord}_G(g)$.*

Theorem A.13 (Euler). *For every n and every $a \in \mathbb{Z}_n^*$, that is, for every pair a, n such that $\gcd(a,n) = 1$, we have $a^{\phi(n)} \equiv 1 \pmod{n}$.*

Proof. First it is easy to check that $(\mathbb{Z}_n^*, \cdot)$ is a group. Then by definition $\phi(n) = |\mathbb{Z}_n^*|$, and since $\langle a \rangle \leq \mathbb{Z}_n^*$, it follows by Lagrange's theorem that $\mathrm{ord}(a) = |\langle a \rangle|$ divides $\phi(n)$. □

Note that Fermat's Little theorem (already presented as theorem A.6) is an immediate consequence of Euler's theorem, since when p is a prime, $\mathbb{Z}_p^* = \mathbb{Z}_p - \{0\}$, and $\phi(p) = (p-1)$.

Theorem A.14 (Chinese Remainder). *Given two sets of numbers of equal size, $r_0, r_1, \ldots, r_n$, and $m_0, m_1, \ldots, m_n$, such that*

$$0 \leq r_i < m_i \qquad 0 \leq i \leq n \tag{A.3}$$

and $\gcd(m_i, m_j) = 1$ for $i \neq j$, then there exists an r such that $r \equiv r_i \pmod{m_i}$ for $0 \leq i \leq n$.

Proof. The proof we give is by counting; we show that the distinct values of r, $0 \leq r < \Pi m_i$, represent distinct sequences. To see that, note that if $r \equiv r' \pmod{m_i}$ for all i, then $m_i | (r - r')$ for all i, and so $(\Pi m_i) | (r - r')$, since the m_i's are pairwise co-prime. So $r \equiv r' \pmod{(\Pi m_i)}$, and so $r = r'$ since both $r, r' \in \{0, 1, \ldots, (\Pi m_i) - 1\}$.

But the total number of sequences $r_0, \ldots, r_n$ such that (A.3) holds is precisely Πm_i. Hence every such sequence must be a sequence of remainders of some r, $0 \leq r < \Pi m_i$. □

Problem A.15. *The proof of theorem A.14 (CRT) is non-constructive. Show how to obtain efficiently the r that meets the requirement of the theorem, i.e., in polytime in n—so in particular not using brute force search.*

Given two groups $(G_1, *_1)$ and $(G_2, *_2)$, a mapping $h : G_1 \longrightarrow G_2$ is a *homomorphism* if it respects the operation of the groups; formally, for all $g_1, g_1' \in G_1$, $h(g_1 *_1 g_1') = h(g_1) *_2 h(g_1')$. If the homomorphism h is also a bijection, then it is called an *isomorphism*. If there exists an isomorphism between two groups G_1 and G_2, we call them *isomorphic*, and write $G_1 \cong G_2$.

If $(G_1, *_1)$ and $(G_2, *_2)$ are two groups, then their product, denoted $(G_1 \times G_2, *)$ is simply $\{(g_1, g_2) : g_1 \in G_1, g_2 \in G_2\}$, where $(g_1, g_2) * (g_1', g_2')$ is $(g_1 *_1 g_1', g_2 *_2 g_2')$. The product of n groups, $G_1 \times G_2 \times \cdots \times G_n$ can be defined analogously; using this notation, the CRT can be stated in the language of group theory as follows.

Theorem A.16 (Chinese Remainder Version II). *If $m_0, m_1, \ldots, m_n$ are pairwise co-prime integers, then $\mathbb{Z}_{m_0 \cdot m_1 \cdot \ldots \cdot m_n} \cong \mathbb{Z}_{m_0} \times \mathbb{Z}_{m_1} \times \cdots \times \mathbb{Z}_{m_n}$.*

Problem A.17. *Prove theorem A.16*

A.1 Answers to selected problems

Problem A.7. $(a+1)^p \equiv_p \sum_{j=0}^{p} \binom{p}{j} a^{p-j} 1^j \equiv_p (a^p + 1) + \sum_{j=1}^{p-1} \binom{p}{j} a^{p-j}$. Note that $\binom{p}{j}$ is divisible by p for $1 \le j \le p-1$, and so we have that $\sum_{j=1}^{p-1} \binom{p}{j} a^{p-j} \equiv_p 0$. Thus we can prove our claim by induction on a. The case $a = 1$ is trivial, and for the induction step we use the above observation to conclude that $(a+1)^p \equiv_p (a^p + 1)$ and we apply the induction hypothesis to get $a^p \equiv_p a$. Once we have prove $a^p \equiv_p a$ we are done since for a such that $\gcd(a, p) = 1$ we have an inverse a^{-1}, so we multiply both sides by it to obtain $a^{p-1} \equiv_p 1$.

Problem A.15. Construct the r in stages, so that at stage i it meets the first i congruences, that is, at stage i we have that $r \equiv r_j \pmod{m_j}$ for $j \in \{0, 1, \ldots, i\}$. Stage 1 is simple: just set $r \longleftarrow r_0$. Suppose that the first i stages have been completed; let $r \longleftarrow r + (\Pi_{j=0}^{i} m_j)x$, where x satisfies

$$x \equiv (\Pi_{j=0}^{i} m_j)^{-1}(r_{i+1} - r) \pmod{m_{i+1}}.$$

We know that the inverse of $(\Pi_{j=0}^{i} m_j)$ exists (in $\mathbb{Z}_{m_{i+1}}$) since $\gcd(m_{i+1}, (\Pi_{j=0}^{i} m_i)) = 1$, and furthermore, this inverse can be obtained efficiently with the extended Euclid's algorithm.

A.2 Notes

For more algebraic background, see [Dummit and Foote (1991)] or [Alperin and Bell (1995)]. For number theory, especially related to cryptography, see [Hoffstein *et al.* (2008)]. A classical text in number theory is [Hardy and Wright (1980)].

Appendix B

Relations

In this section we present the basics of relations. Given two sets X, Y, $X \times Y$ denotes the set of (ordered) pairs $\{(x,y)|x \in X \wedge y \in Y\}$, and a *relation* R is just a subset of $X \times Y$, i.e., $R \subseteq X \times Y$. Thus, the elements of R are of the form (x,y) and we write $(x,y) \in R$ (we can also write xRy, Rxy or $R(x,y)$). In what follows we assume that we quantify over the set X and that $R \subseteq X \times X$; we say that

(1) R is *reflexive* if $\forall x$, $(x,x) \in R$,
(2) R is *symmetric* if $\forall x \forall y$, $(x,y) \in R$ if and only if $(y,x) \in R$,
(3) R is *antisymmetric* if $\forall x \forall y$, if $(x,y) \in R$ and $(y,x) \in R$ then $x = y$,
(4) R is *transitive* if $\forall x \forall y \forall z$, if $(x,y) \in R$ and $(y,z) \in R$ then it is also the case that $(x,z) \in R$.

Suppose that $R \subseteq X \times Y$ and $S \subseteq Y \times Z$. The *composition* of R and S is defined as follows:

$$R \circ S = \{(x,y)|\exists z, xRz \wedge zSy\}. \tag{B.1}$$

Let $R \subseteq X \times X$; we can define $R^n := R \circ R \circ \cdots \circ R$ recursively as follows:

$$R^0 = \mathrm{id}_X := \{(x,x)|x \in X\}, \tag{B.2}$$

and $R^{i+1} = R^i \circ R$. Note that there are two different equalities in (B.2); "=" is the usual equality, and ":=" is a definition.

Theorem B.1. *The following three are equivalent:*

(1) R is transitive,
(2) $R^2 \subseteq R$,
(3) $\forall n \geq 1$, $R^n \subseteq R$.

Problem B.2. *Prove theorem B.1.*

There are two standard ways of representing *finite* relations, that is, relations on $X \times Y$ where X and Y are finite sets. Let $X = \{a_1, \ldots, a_n\}$ and $Y = \{b_1, \ldots, b_m\}$, then we can represent a relation $R \subseteq X \times Y$:

(1) as a matrix $M_R = (m_{ij})$ where:

$$m_{ij} = \begin{cases} 1 & (a_i, b_j) \in R \\ 0 & (a_i, b_j) \notin R \end{cases},$$

(2) and as a directed graph $G_R = (V_R, E_R)$, where $V_R = X \cup Y$ and ${}_{a_i}\bullet \longrightarrow \bullet_{b_j}$ is an edge in E_R iff $(a_i, b_j) \in R$.

B.1 Closure

Let P be a property[4] of relations, for example transitivity or symmetry. Let $R \subseteq X \times X$ be a relation, with or without the property P. The relation S satisfying the following three conditions:

$$\begin{array}{l} (1)\ S \text{ has the property P} \\ (2)\ R \subseteq S \\ (3)\ \forall Q \subseteq X \times X, \text{ “}Q \text{ has P” and } R \subseteq Q \text{ implies that } S \subseteq Q \end{array} \tag{B.3}$$

is called the *closure of R with respect to* P. Note that in some instances the closure may not exist. Also note that condition 3 may be replaced by

$$S \subseteq \bigcap_{Q \text{ has P},\ R \subseteq Q} Q. \tag{B.4}$$

See figure B.1 for an example of reflexive closure.

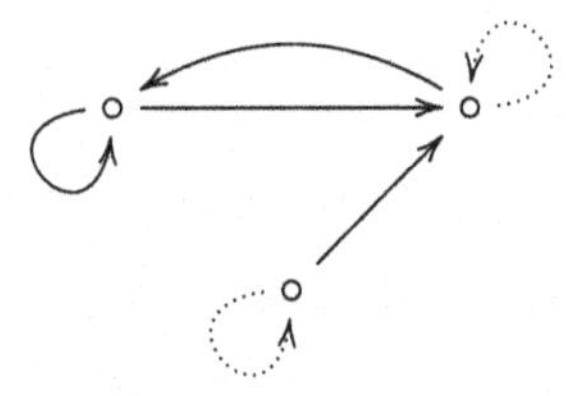

Fig. B.1 Example of reflexive closure: without the dotted lines, this diagram represents a relation that is not reflexive; with the dotted lines it is reflexive, and it is in fact the smallest reflexive relation containing the three points and four solid lines.

[4]We have seen the concept of an abstract property in section 1.1. The only difference is that in section 1.1 the property $P(i)$ was over $i \in \mathbb{N}$, whereas here, given a set X, the property is over $Q \in \mathcal{P}(X \times X)$, that is, $P(Q)$ where $Q \subseteq X \times X$. In this section, instead of writing $P(Q)$ we say “Q has property P.”

Theorem B.3. *For $R \subseteq X \times X$, $R \cup \mathrm{id}_X$ is the reflexive closure of R.*

Problem B.4. *Prove theorem B.3.*

See figure B.2 for an example of symmetric closure.

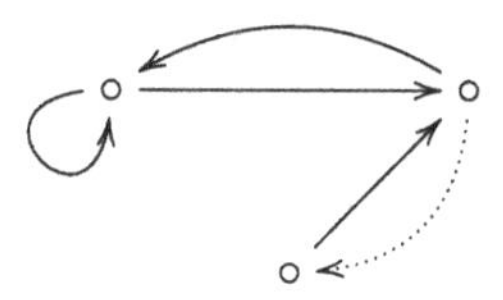

Fig. B.2 Example of symmetric closure: without the dotted line, this diagram represents a relation that is not symmetric; with the dotted lines it is symmetric.

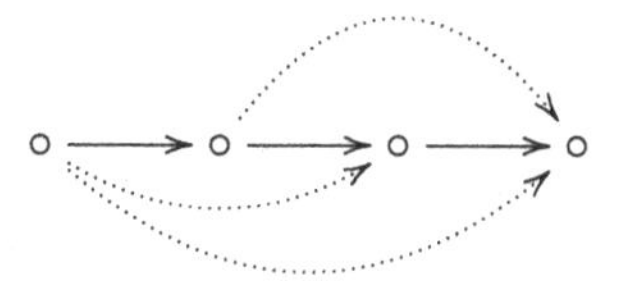

Fig. B.3 Example of transitive closure: without the dotted line, this diagram represents a relation that is not transitive; with the dotted lines it is transitive.

Theorem B.5. *Given a relation $R \subseteq X \times Y$, the relation $R^{-1} \subseteq Y \times X$ is defined as $\{(x, y)|(y, x) \in R\}$. For $R \subseteq X \times X$, $R \cup R^{-1}$ is the symmetric closure of R.*

Problem B.6. *Prove theorem B.5.*

See figure B.3 for an example of transitive closure.

Theorem B.7. $R^+ := \bigcup_{i=1}^{\infty} R^i$ *is the transitive closure of R.*

Proof. We check that R^+ has the three conditions given in (B.3). First, we check whether R^+ has the given property, i.e., whether it is transitive:

$$\begin{aligned} xR^+y \wedge yR^+z &\iff \exists m, n \geq 1, xR^m y \wedge yR^n z \\ &\implies \exists m, n \geq 1, x(R^m \circ R^n)z \qquad (\dagger) \\ &\iff \exists m, n \geq 1, xR^{m+n}z \\ &\iff xR^+z \end{aligned}$$

so R^+ *is* transitive.

Second we check that $R \subseteq R^+$, but this follows by the definition of R^+.

We check now the last condition. Suppose S is transitive and $R \subseteq S$. Since S is transitive, by theorem B.1, $S^n \subseteq S$, for $n \geq 1$, i.e., $S^+ \subseteq S$, and since $R \subseteq S$, $R^+ \subseteq S^+$, so $R^+ \subseteq S$. □

Problem B.8. *Note that in the proof of theorem B.7, when we show that R^+ itself is transitive, the second line, labeled with (†), is an implication, rather than an equivalence like the other lines. Why is it not an equivalence?*

Theorem B.9. $R^* = \bigcup_{i=0}^{\infty} R^i$ *is the reflexive and transitive closure of* R.

Proof. $R^* = R^+ \cup \mathrm{id}_X$. □

B.2 Equivalence relation

Let X be a set, and let I be an index set. The family of sets $\{A_i | i \in I\}$ is called a *partition* of X iff

(1) $\forall i,\ A_i \neq \emptyset$,
(2) $\forall i \neq j,\ A_i \cap A_j = \emptyset$,
(3) $X = \bigcup_{i \in I} A_i$.

Note that $X = \bigcup_{x \in X} \{x\}$ is the *finest* partition possible, i.e., the set of all singletons. A relation $R \subseteq X \times X$ is called an *equivalence relation* iff

(1) R is reflexive,
(2) R is symmetric,
(3) R is transitive.

For example, if x, y are strings over $\{0,1\}^*$, then the relation given by $R = \{(x,y) | \mathrm{length}(x) = \mathrm{length}(y)\}$ is an equivalence relation. Another example is $xRy \iff x = y$, i.e., the equality relation is the equivalence relation *par excellence*. Yet another example: $R = \{(a,b) | a \equiv b \pmod{m}\}$ is an equivalence relation (where "$\equiv$" is the congruence relation defined on page 152).

Theorem B.10. *Let $F : X \longrightarrow X$ be any* total *function (i.e., a function defined on all its inputs). Then the relation R on X defined as: $xRy \iff F(x) = F(y)$, is an equivalence relation.*

Problem B.11. *Prove theorem B.10.*

Let R be an equivalence relation on X. For every $x \in X$, the set $[x]_R = \{y | xRy\}$ is the *equivalence class of x with respect to R*.

Theorem B.12. *Let $R \subseteq X \times X$ be an equivalence relation. The following are equivalent:*

(1) aRb
(2) $[a] = [b]$
(3) $[a] \cap [b] \neq \emptyset$

Proof. (1) $\Rightarrow$ (2) Suppose that aRb, and let $c \in [a]$. Then aRc, so cRa (by symmetry). Since $cRa \wedge aRb$, cRb (transitivity), so bRc (symmetry), so $c \in [b]$. Hence $[a] \subseteq [b]$, and similarly $[b] \subseteq [a]$.

(2) $\Rightarrow$ (3) Obvious, since $[a]$ is non-empty as $a \in [a]$.

(3) $\Rightarrow$ (1) Let $c \in [a] \cap [b]$, so aRc and bRc, so by symmetry $aRc \wedge cRb$, so by transitivity aRb. □

Corollary B.13. *If R is an equivalence, then $(a, b) \notin R$ iff $[a] \cap [b] = \emptyset$.*

For every equivalence relation $R \subseteq X \times X$, let X/R denote the set of all equivalence classes of R.

Theorem B.14. *X/R is a partition of X.*

Proof. Given theorem B.12, the only thing that remains to be proven is that $X = \bigcup_{A \in X/R} A$. Since every $A = [a]$ for some $a \in X$, it follows that $\bigcup_{A \in X/R} A = \bigcup_{a \in X} [a] = X$. □

Let R_1, R_2 be equivalence relations. If $R_1 \subseteq R_2$, then we say that R_1 is a *refinement* of R_2.

Lemma B.15. *If R_1 is a refinement of R_2, then $[a]_{R_1} \subseteq [a]_{R_2}$, for all $a \in X$.*

If X/R is finite then $\text{index}(R) := |X/R|$, i.e., the *index* of R (in X) is the size of X/R.

Theorem B.16. *If $R_1 \subseteq R_2$, then* $\text{index}(R_1) \geq \text{index}(R_2)$.

Problem B.17. *Prove theorem B.16.*

B.3 Partial orders

In this section, instead of using R to represent a relation over a set X, we are going to use the different variants of inequality: $(X, \preceq), (X, \sqsubseteq), (X, \leq)$.

A relation $\preceq$ over X (where $\preceq \subseteq X \times X$) is called a *partial order* (a *poset* for short) if it is

(1) reflexive
(2) antisymmetric
(3) transitive

A relation "$\prec$" (where $\prec \subseteq X \times X$) is a *sharp partial order* if

(1) $x \prec y \Rightarrow \neg(y \prec x)$
(2) transitive

These two standard relations, "$\preceq$" and "$\prec$", are linked in a natural manner by the following theorem.

Theorem B.18. *A relation $\preceq$ defined as $x \preceq y \iff x \prec y \vee x = y$ is a partial order. That is, given a sharp partial order "$\prec$", we can extend it to a poset "$\preceq$" with the standard equality symbol "$=$".*

Let $(X, \preceq)$ be a poset. We say that x, y are *comparable* if $x \preceq y$ or $y \preceq x$. Otherwise, they are *incomparable*. Let $x \sim y$ be short for x, y are incomparable, i.e., $x \sim y \iff \neg(x \preceq y) \wedge \neg(y \preceq x)$. In general, for every pair x, y exactly one of the following is true

$$x \prec y, \quad y \prec x, \quad x = y, \quad x \sim y$$

Of course, in the context of posets represented by "$\preceq$" the meaning of "$\prec$" is as follows: $x \prec y \iff x \preceq y \wedge x \neq y$.

A poset $(X, \preceq)$ is *total* or *linear* if all x, y are comparable, i.e., $\sim = \emptyset$.

Some examples of posets: if X is a set, then $(\mathcal{P}(X), \subseteq)$ is a poset. For example, if $X = \{1, 2, 3\}$, then a *Hasse diagram* representation of this poset would be as given in figure B.4.

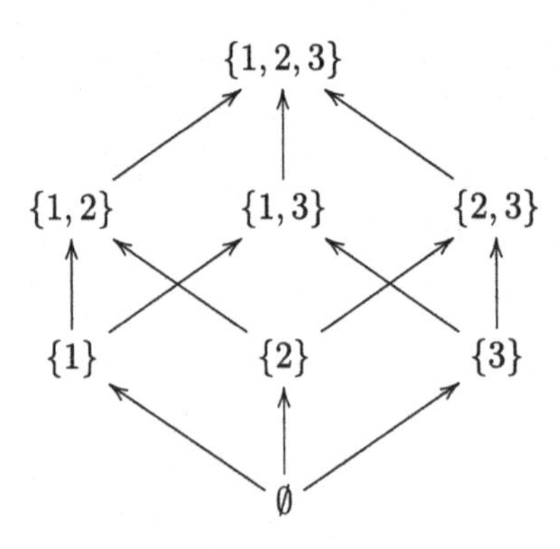

Fig. B.4 Hasse diagram representation of the poset $(\{1, 2, 3\}, \subseteq)$.

Let $\mathbb{Z}^+$ be the set of positive integers, and let $a|b$ be the "a divides b" relation (that we define on page 151). Then, $(\mathbb{Z}^+, |)$ is a poset.

If $(X_1, \preceq_1), (X_2, \preceq_2)$ are two posets, then the *component-wise order* is $(X_1 \times X_2, \preceq_C)$ defined as follows:

$$(x_1, x_2) \preceq (y_1, y_2) \iff x_1 \preceq_1 y_1 \wedge x_2 \preceq_2 y_2,$$

and it is also a poset.

The *lexicographic order* $(X_1 \times X_2, \preceq_L)$ is defined as follows:

$$(x_1, x_2) \preceq_L (y_1, y_2) \iff (x_1 \preceq_1 y_1) \vee (x_1 = y_1 \wedge x_2 \preceq_2 y_2).$$

Finally, $(X, \preceq)$ is a *stratified order* iff $(X, \preceq)$ is a poset, and furthermore $(x \sim y \wedge y \sim z) \Rightarrow (x \sim z \vee x = z)$. Define $a \approx b \iff a \sim b \vee a = b$.

Theorem B.19. *A poset $(X, \preceq)$ is a stratified order iff $\approx = \sim \cup \operatorname{id}_X$ is an equivalence relation.*

In mathematics nomenclature can be the readers greatest scourge. The string of symbols "$\approx = \sim \cup \operatorname{id}_X$" is a great example of obfuscation; how to make sense of it? Yes, it is very succinct, but it takes practice to be able to read it. What we are saying here is that the order we called "$\approx$" is actually equal to the order that we obtain by taking the union of the order "$\sim$" and "id_X".

Problem B.20. *Prove theorem B.19.*

Theorem B.21. *A poset $(X, \preceq)$ is a stratified order iff there exists a total order $(T, \preceq_T)$ and an function $f : X \longrightarrow T$ such that f is onto and f is an "order homomorphism," i.e., $a \preceq b \iff f(a) \preceq_T f(b)$.*

Problem B.22. *Prove theorem B.21.*

B.4 Lattices

Let $(X, \preceq)$ be a poset, and let $A \subseteq X$ be a subset, and $a \in X$. Then:

(1) a is *minimal* in X if $\forall x \in X, \neg(x \prec a)$.
(2) a is *maximal* in X if $\forall x \in X, \neg(a \prec x)$.
(3) a is the *least element* in X if $\forall x \in X$, $a \preceq x$.
(4) a is the *greatest element* in X if $\forall x \in X, x \preceq a$.
(5) a is an *upper bound* of A if $\forall x \in A, x \preceq a$.
(6) a is a *lower bound* if A if $\forall x \in A, a \preceq x$.

(7) a is the *least upper bound (supremum)* of A, denoted $\sup(A)$ if
   (a) $\forall x \in A, x \preceq a$
   (b) $\forall b \in X, (\forall x \in A, x \preceq b) \Rightarrow a \preceq b$
(8) a is the *greatest lower bound (infimum)* of A, denoted $\inf(A)$ if
   (a) $\forall x \in A, a \preceq x$
   (b) $\forall b \in X, (\forall x \in A, b \preceq x) \Rightarrow b \preceq a$

Problem B.23. *Note that in the definitions 1–8 we sometimes use the definite article "the" and sometimes the indefinite article "a". In the former case this implies* uniqueness; *in the latter case this implies that there may be several candidates. Convince yourself of uniqueness where it applies, and provide an example of a poset where there are several candidates for a given element in the other cases. Finally, it is important to note that* $\sup(A), \inf(A)$ *may or may not exist; provide examples where they do not exist.*

A poset $(X, \preceq)$ is a *well-ordered* set if it is a total order and for every $A \subseteq X$, such that $A \neq \emptyset$, A has a least element.

A poset is *dense* if $\forall x, y$ if $x < y$, then $\exists z, x < z < y$. For example, $(\mathbb{R}, \leq)$, with a standard definition of "$\leq$", is a total dense order, but it is not a well ordered set; for example, the interval $(2, 3]$, which equals the subset of $\mathbb{R}$ consisting of those x such that $2 < x \leq 3$, does not have a least element.

A poset $(X, \preceq)$ is a *lattice* if $\forall a, b \in X$, $\inf(\{a, b\})$ and $\sup(\{a, b\})$ both exist in X. For example, every total order is a lattice, and $(\mathcal{P}(X), \subseteq)$ is a lattice for every X. This last example inspires the following notation: $a \sqcup b := \sup(\{a, b\})$ and $a \sqcap b := \inf(\{a, b\})$.

Problem B.24. *Prove that for the lattice* $(\mathcal{P}(X), \subseteq)$ *we have:*

$$A \sqcup B = A \cup B$$
$$A \sqcap B = A \cap B$$

Not every poset is a lattice; figure B.5 gives an easy example.

Theorem B.25. *Let* $(X, \preceq)$ *be a lattice. Then,* $\forall a, b \in X$,

$$a \preceq b \iff a \sqcap b = a \iff a \sqcup b = b.$$

Problem B.26. *Prove theorem B.25.*

Theorem B.27. *Let* $(X, \preceq)$ *be a lattice. Then, the following hold for all* $a, b, c \in X$*:*

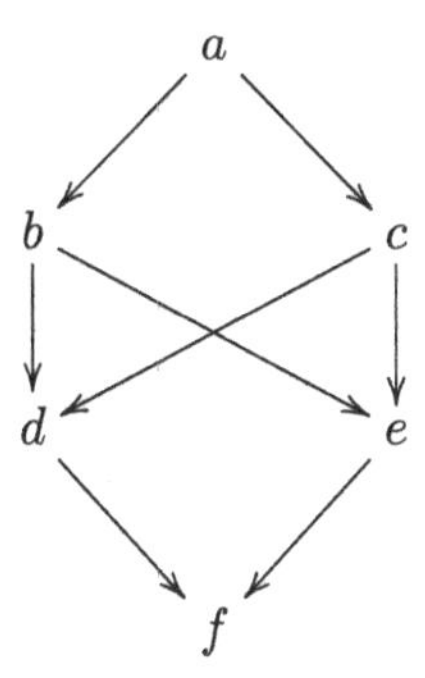

Fig. B.5 An example of a poset that is *not* a lattice. While $\inf(\{b,c\}) = a$ and $\sup(\{d,e\}) = f$, the supremum of $\{b,c\}$ does not exist.

(1) $a \sqcup b = b \sqcup a$ *and* $a \sqcap b = b \sqcap a$ *(commutativity)*
(2) $a \sqcup (b \sqcup c) = (a \sqcup b) \sqcup c$ *and* $a \sqcap (b \sqcap c) = (a \sqcap b) \sqcap c$ *(associativity)*
(3) $a \sqcup a = a$ *and* $a \sqcap a = a$ *(idempotence)*
(4) $a = a \sqcup (a \sqcap b)$ *and* $a = a \sqcap (a \sqcup b)$ *(absorption)*

Problem B.28. *Prove the properties listed as theorem B.27.*

A lattice $(X, \preceq)$ is *complete* iff $\forall A \subseteq X$, $\sup(A), \inf(A)$ both exist. We denote $\bot = \inf(X)$ and $\top = \sup(X)$.

Theorem B.29. *$(\mathcal{P}(X), \subseteq)$ is a complete lattice, and the following hold $\forall \mathcal{A} \subseteq \mathcal{P}(X)$, $\sup(\mathcal{A}) = \bigcup_{A \in \mathcal{A}} A$ and $\inf(\mathcal{A}) = \bigcap_{A \in \mathcal{A}} A$, and $\bot = \emptyset$ and $\top = X$.*

Problem B.30. *Prove theorem B.29.*

Theorem B.31. *Every finite lattice is complete.*

Proof. Let $A = \{a_1, \ldots, a_n\}$. Define $b = a_1 \sqcap \ldots \sqcap a_n$ (with parenthesis associated to the right). Then $b = \inf(A)$. Same idea for the supremum. □

B.5 Fixed point theory

Suppose that F is a function, and consider the equation $\vec{x} = F(\vec{x})$. A solution $\vec{a}$ of this equation is a *fixed point of* F.

Let $(X, \preceq)$ and $(Y, \sqsubseteq)$ be two posets. A function $f : X \longrightarrow Y$ is *monotone* iff $\forall x, y \in X$, $x \preceq y \Rightarrow f(x) \sqsubseteq f(y)$. For example,

$f_B : \mathcal{P}(X) \longrightarrow \mathcal{P}(X)$, where $B \subseteq X$, defined as $\forall x \subseteq X$, $f_B(x) = B - x$ is *not* monotone. On the other hand, $g_B(x) = B \cup x$ and $h_B(x) = B \cap x$ are both monotone.

Let $(X, \preceq)$ be a poset, and let $f : X \longrightarrow X$. A value $x_0 \in X$ such that $x_0 = f(x_0)$ is, as we saw, a fixed point of f. A fixed point may not exist; for example, f_B in the above paragraph does not have a fixed point when $B \neq \emptyset$, since the set equation $x = B - x$ does not have a solution in that case. There may also be many fixed points; for example, $f(x) = x$ has $|X|$ many fixed points.

Theorem B.32 (Knaster-Tarski (1)). *Let $(X, \preceq)$ be a complete lattice, and let $f : X \longrightarrow X$ be a monotone function. Then the least fixed point of the equation $x = f(x)$ exists and it is equal to* $\inf(\{x|f(x) \preceq x\})$.

Proof. Let $x_0 = \inf(\{x|f(x) \preceq x\})$. First we show that $x_0 = f(x_0)$. Let $B = \{x|f(x) \preceq x\}$, and note that $B \neq \emptyset$ because $\top = \sup(X) \in B$. Let $x \in B$, so we have $x_0 \preceq x$, hence since f is monotone, $f(x_0) \preceq f(x)$, i.e.,

$$f(x_0) \preceq f(x) \preceq x.$$

This is true for each x in B, so $f(x_0)$ is a lower bound for B, and since x_0 is the greatest lower bound of B, it follows that $f(x_0) \preceq x_0$.

Since f is monotone it follows that $f(f(x_0)) \preceq f(x_0)$, which means that $f(x_0)$ is in B. But then $x_0 \preceq f(x_0)$, which means that $x_0 = f(x_0)$.

It remains to show that x_0 is the least fixed point. Let $x' = f(x')$. This means that $f(x') \preceq x'$, i.e., $x' \in B$. But then $x_0 \preceq x'$. □

Theorem B.33 (Knaster-Tarski (2)). *Let $(X, \preceq)$ be a complete lattice, and let $f : X \longrightarrow X$ be a monotone function. Then the greatest fixed point of the equation $x = f(x)$ exists and it is equal to* $\sup(\{x|f(x) \preceq x\})$.

Note that these theorems are not constructive, but in the case of finite X, there is a constructive way of finding the least and greatest fixed points.

Theorem B.34 (Knaster-Tarski: finite sets). *Let $(X, \preceq)$ be a lattice, $|X| = m$, $f : X \longrightarrow X$ a monotone function. Then $f^m(\bot)$ is the least fixed point, and $f^m(\top)$ is the greatest fixed point.*

Proof. Since $|X| = m$, $(X, \preceq)$ is a complete lattice, $\bot = \inf(X)$ and $\top = \sup(X)$ both exist. Since f is monotone, and $\bot \preceq f(\bot)$, we have $f(\bot) \preceq f(f(\bot))$, i.e., $f(\bot) \preceq f^2(\bot)$. Continuing to apply monotonicity we obtain:

$$f^0(\bot) = \bot \preceq f(\bot) \preceq f^2(\bot) \preceq f^3(\bot) \preceq \cdots \preceq f^i(\bot) \preceq f^{i+1}(\bot) \preceq \cdots.$$

Consider the above sequence up to $f^m(\bot)$. It has length $(m+1)$, but X has only m elements, so there are $i < j$, such that $f^i(\bot) = f^j(\bot)$. Since $\preceq$ is an order, it follows that

$$f^i(\bot) = f^{i+1}(\bot) = \cdots = f^j(\bot),$$

so $x_0 = f^i(\bot)$ is a fixed point as

$$f(x_0) = f(f^i(\bot)) = f^{i+1}(\bot) = f^i(\bot) = x_0.$$

Clearly $f^{j+1}(\bot) = f(f^j(\bot)) = f(x_0) = x_0$, so in fact $\forall k \geq i$, $x_0 = f^k(\bot)$, and so $f^m(\bot) = x_0$, so $f^m(\bot)$ is a fixed point.

We now suppose that x is another fixed point of f, i.e., $x = f(x)$. Since $\bot \preceq x$, and f is monotone, we conclude $f(\bot) \preceq f(x) = x$, i.e., $f(\bot) \preceq x$. Again, since f is monotone, $f(f(\bot)) \preceq f(x) = x$, so $f^2(\bot) \preceq x$. Hence, repeating this procedure sufficiently many times, we obtain $f^i(\bot) \preceq x$ for each i, so we get $x_0 = f^m(\bot) \preceq x$.

We do a similar argument for "greatest." □

The situation is even better for the standard lattice $(\mathcal{P}(X), \subseteq)$, if X is finite.

Theorem B.35. *Let X be a finite set, $|X| = n$, $f : \mathcal{P}(X) \longrightarrow \mathcal{P}(X)$ is monotone. Then $f^{n+1}(\emptyset)$ is the least fixed point, and $f^{n+1}(X)$ is the greatest fixed point.*

Proof. Note that the previous theorem says that $f^{2^n}(\emptyset)$ is the least fixed point, and $f^{2^n}(X)$ is the greatest fixed point, since $|\mathcal{P}(X)| = 2^n$, $\bot = \emptyset$ and $\top = X$, for the lattice $(\mathcal{P}(X), \subseteq)$. But this theorem claims $(n+1)$ instead of 2^n. The reason is that $\emptyset \subseteq f(\emptyset) \subseteq f^2(\emptyset) \subseteq \cdots \subseteq f^{n+1}(\emptyset)$ *must* have two repeating sets (because $|X| = n$). □

Problem B.36. *Consider the lattice $(\mathcal{P}(\{a,b,c\}), \subseteq)$ and the functions $f(x) = x \cup \{a,b\}$ and $g(x) = x \cap \{a,b\}$. Compute their respective least/greatest fixed points.*

Let $(X, \preceq)$ be a complete lattice. A function $f : X \longrightarrow X$ is called

(1) *upward continuous* iff $\forall A \subseteq X$, $f(\sup(A)) = \sup(f(A))$,
(2) *downward continuous* iff $\forall A \subseteq X$, $f(\inf(A)) = \inf(f(A))$,
(3) *continuous* if it is both upward and downward continuous.

Lemma B.37. *If $f : X \longrightarrow X$ is upward (downward) continuous, then it is monotone.*

Proof. Let f be upward continuous and $x \preceq y$, so $x = \inf(\{x, y\})$ and $y = \sup(\{x, y\})$, and

$$f(x) \preceq \sup(\{f(x), f(y)\}) = \sup(f(\{x, y\})) = f(\sup(\{x, y\})) = f(y).$$

A similar argument for downward continuous. □

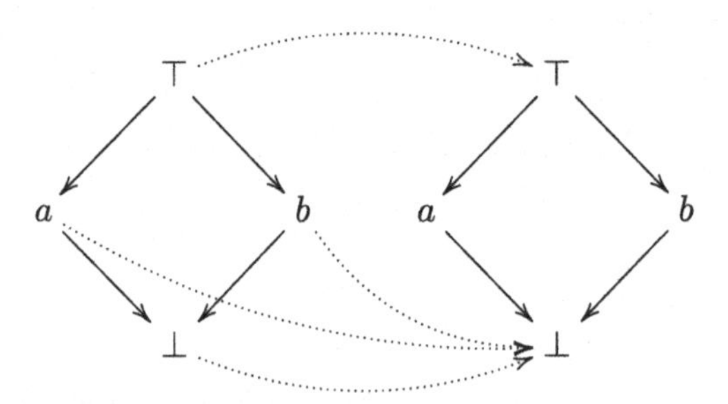

Fig. B.6 An example of an ordering over $X = \{a, b, \bot, \top\}$, with a function $f : X \longrightarrow X$, indicated by the dotted lines. That is, $f(\top) = \top$ and $f(a) = f(b) = f(\bot) = \bot$. It can be checked by inspection that f is monotone, but it is not upward continuous.

Problem B.38. *Show that the function f in figure B.6 is not upward continuous. Give an example of a monotone function g that is neither upward nor downward continuous.*

Theorem B.39 (Kleene). *If $(X, \preceq)$ is a complete lattice, $f : X \longrightarrow X$ is an upward continuous function, then $x_0 = \sup(\{f^n(\bot)|n = 1, 2, \ldots\})$ is the least fixed point of f.*

Proof. Note that $\bot \preceq f(\bot)$, so by monotonicity of f, we have that

$$\bot \preceq f(\bot) \preceq f^2(\bot) \preceq f^3(\bot) \preceq \cdots \tag{B.5}$$

and,

$$f(x_0) = f(\sup(\{f^n(\bot)|n = 1, 2, \ldots\}))$$

and since f is upward continuous

$$= \sup(f(\{f^n(\bot)|n = 1, 2, \ldots\}))$$
$$= \sup(\{f^{n+1}(\bot)|n = 1, 2, \ldots\})$$

and by (B.5),

$$= \sup(\{f^n(\bot) | n = 1, 2, \ldots\}) = x_0$$

so $f(x_0) = x_0$, i.e., x_0 is a fixed point.

Let $x = f(x)$. We have $\bot \preceq x$ and f is monotone, so $f(\bot) \preceq f(x) = x$, i.e., $f(\bot) \preceq x$, $f^2(\bot) \preceq f(x) = x$, etc., i.e., $f^n(\bot) \preceq x$, for all n, so by the definition of sup,

$$x_0 = \sup(\{f^n(\bot) | n = 1, 2, \ldots\}) \preceq x,$$

so x_0 is the least fixed point. □

B.6 Answers to selected problems

Problem B.2. (1) ⇒ (2) Suppose that R is transitive, and let $(x, y) \in R^2$. Then, by definition (B.1) we know that there exists a z such that xRz and zRy. By transitivity we have that $(x, y) \in R$. (2) ⇒ (3) Suppose that $R^2 \subseteq R$. We show by induction on n that $R^n \subseteq R$. The basis case, $n = 1$, is trivial. For the induction step, suppose that $(x, y) \in R^{n+1} = R^n \circ R$, so by definition (B.1) there exists a z such that xR^nz and zRy. By the induction assumption this means that xRz and zRy, so $(x, y) \in R^2$, and since $R^2 \subseteq R$, it follows that $(x, y) \in R$, and we are done. (3) ⇒ (1) Suppose that for all n, $R^n \subseteq R$. If xRy and yRz then $xRz \in R^2$, and so $xRz \in R$, and so R is transitive.

Problem B.8. The reason is that in the first line we chose a *particular* y: $xR^+y \wedge yR^+z \iff \exists m, n \geq 1, xR^my \wedge yR^nz$. On the other hand, from the statement $\exists m, n \geq 1, x(R^m \circ R^n)z$ we can only conclude that there exists some y' such that $\exists m, n \geq 1, xR^my' \wedge y'R^nz$, and it is not necessarily the case that $y = y'$.

Problem B.11. R is reflexive since $F(x) = F(x)$; R is symmetric since $F(x) = F(y)$ implies $F(y) = F(x)$ (equality is a symmetric relation); R is transitive because $F(x) = F(y)$ and $F(y) = F(z)$ implies $F(x) = F(z)$ (again by transitivity of equality).

Problem B.17. We know from lemma B.15 that $\forall a \in X$, $[a]_{R_1} \subseteq [a]_{R_2}$. Therefore the mapping $f : X/R_1 \longrightarrow X/R_2$ given by $f([a]_{R_1}) = [a]_{R_2}$ is surjective, and hence $|X/R_1| \geq |X/R_2|$.

Problem B.20. We show the left-to-right direction. Clearly $\approx$ is reflexive as it contains id_X. Now suppose that $a \approx b$; then $a \sim b$ or $a = b$. If $a = b$, then $b = a$ (as equality is obviously a symmetric relation), and so $b \approx a$. If $a \sim b$, then by definition of incomparability, $\neg(a \preceq b) \wedge \neg(b \preceq a)$, which is logically equivalent to $\neg(b \preceq a) \wedge \neg(a \preceq b)$, and hence $b \sim a$, and so $b \approx a$ in this case as well. Finally, we want to prove transitivity: suppose that $a \approx b \wedge b \approx c$; if $a = b$ and $b = c$, then $a = c$ and we have $a \approx c$. Similarly, if $a = b$ and $b \sim c$, then $a \sim c$, and so $a \approx c$, and if $a \sim b$ and $b = c$, and $a \sim c$, and also $a \approx c$. The only case that remains is $a \sim b$ and $b \sim c$, and it is here where we use the fact that $\preceq$ is a stratified order, as this implies that $a \sim c \vee a = c$, which gives us $a \approx c$.

Problem B.22. We show the left-to-right direction. The natural way to proceed here is to let T be the set consisting of the different equivalence classes of X under $\sim$. That is, $T = \{[a]_\sim : a \in X\}$. Then T is totally ordered under $\preceq_T$ defined as follows: for $X, X' \in T$, such that $X \neq X'$ and $X = [x]$ and $X' = [x']$, we have that $X \preceq_T X'$ iff $x \preceq x'$. Note also that given two distinct X, X' in T, and any pair of representatives x, x', it is always the case that $x \preceq x'$ or $x' \preceq x$, since if neither was the case, we would have $x \sim x'$, and hence $[x] = [x']$ and so $X = X'$. Then the function $f : X \longrightarrow T$ given as $f(x) = [x]$ satisfies the requirements.

Problem B.26. We prove the following part: $a \preceq b \iff a \sqcap b = a$. Suppose that $a \preceq b$. As $(X, \preceq)$ is a lattice, it is a poset, and so $a \preceq a$ (reflexivity), which means that a is a lower bound of the set $\{a, b\}$. Since $(X, \preceq)$ is a lattice, $\inf\{a, b\}$ exists, and thus $a \preceq \inf\{a, b\}$. On the other hand, $\inf\{a, b\} \preceq a$, and so, by the antisymmetry of a poset, we have $a = \inf\{a, b\} = a \sqcap b$. For the other direction, $a \sqcap b = a$ means that $\inf\{a, b\} = a$, and so $a \preceq \inf\{a, b\}$, and so $a \preceq b$.

Problem B.28. (1) follows directly from the observation that $\{a, b\}$ and $\{b, a\}$ are the same set. (2) follows from the observation that $\inf\{a, \inf\{b, c\}\} = \inf\{a, b, c\} = \inf\{\inf\{a, b\}, c\}$, and same for the supremum. (3) follows directly from the observation that $\{a, a\} = \{a\}$ (we are dealing with sets, not with "multi-sets"). For (4), the absorption law, we show that $a = a \sqcup (a \sqcap b)$. First note that $a \preceq \sup\{a, *\}$ (where "$*$" denotes anything, in particular $a \sqcap b$). On the other hand, $a \sqcap b \preceq a$ by definition, and $a \preceq a$ by reflexivity, and so a is upper bound for the set $\{a, a \sqcap b\}$. Therefore, $\sup\{a, a \sqcap b\} \preceq a$ and hence, by antisymmetry, $a = \sup\{a, a \sqcap b\}$, i.e., $a = a \sqcup (a \sqcap b)$. The other absorption law can be proven similarly.

Problem B.36. For example, the least fixed point of f is given by $f^4(\emptyset) = f^3(\{a, b\}) = f^2(\{a, b\}) = f(\{a, b\}) = \{a, b\}$.

Problem B.38. Note that $\sup\{a, b\} = \top$, and so $f(\sup\{a, b\}) = f(\top) = \top$. On the other hand, $f(\{a, b\}) = \{\bot\}$, as $f(a) = f(b) = \bot$. Therefore, $\sup(f(\{a, b\}) = \sup(\{\bot\}) = \bot$. See figure B.7 for a function g that is monotone, but is neither upward nor downward continuous.

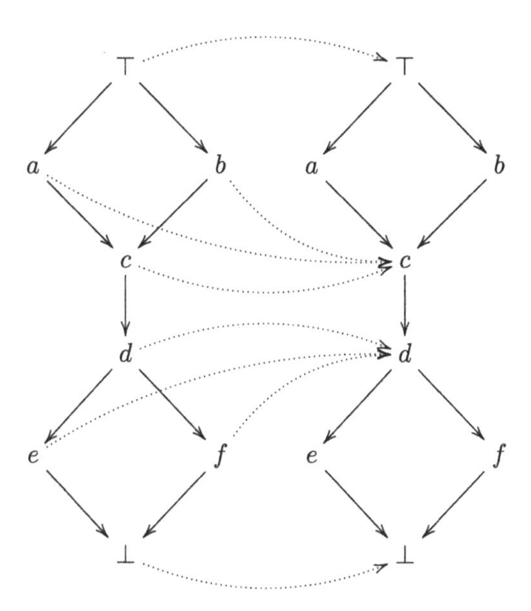

Fig. B.7 An example of an ordering over $X = \{a, b, c, d, e, f, \bot, \top\}$, with a function $g : X \longrightarrow X$, indicated by the dotted lines. While g is monotone, it is neither upward now downward continuous.

B.7 Notes

This chapter is based on hand-written lecture slides of Ryszard Janicki. A basic introduction to relations can be found in chapter 8 of [Rosen (2007)], and for a very quick introduction to relations (up to the definition of equivalence classes), the reader is invited to read the delightful section 7 of [Halmos (1960)].

A different perspective on partial orders is offered in [Mendelson (1970)], chapter 3. In this book the author approaches partial orders from the point of view of *Boolean algebras*, which are defined as follows: a set B together with two binary operations $\curlywedge, \curlyvee$ (normally denoted $\wedge, \vee$, but we use these for "and","or", and so here we use the "curly" version to emphasize that "$\curlywedge$" and "$\curlyvee$" are not necessarily the standard Boolean connectives) on B, a singularity operation $'$ on B, and two specific elements 0 and 1 of B, and satisfying a set of axioms: $x \curlywedge y = y \curlywedge x$ and $x \curlyvee y = y \curlyvee x$, distributivity of $\curlywedge$ over $\curlyvee$, and vice-versa, as well as $x \curlywedge 1 = x$ and $x \curlyvee 0 = x$, $x \curlyvee x' = 1$ and $x \curlywedge x' = 0$, and finally $0 \neq 1$. A Boolean algebra is usually denoted by the sextuple $\mathcal{B} = \langle B, \curlywedge, \curlyvee, ', 0, 1\rangle$, and it is assumed to satisfy the axioms just listed.

Given a Boolean algebra $\mathcal{B}$, we define a binary relation $\preceq$ as follows:

$$x \preceq y \iff x \curlywedge y = x.$$

This turns out to be equivalent to our notion of a lattice order. Mendelson then abstracts the three properties of reflexivity, antisymmetry and transitivity, and says that any relation that satisfies all three is a partial order—and not every partial order is a lattice.

Appendix C

Logic

We present the foundations of propositional and predicate logic with the aim of defining Peano Arithmetic (PA). PA is the standard formalization of number theory, and it is the logical background for section 1.3.6—Formal Verification. Our treatment of logic is limited to providing this background, but the reader can find more resources in the Notes section.

C.1 Propositional Logic

Propositional (Boolean) formulas are built from *propositional (Boolean) variables*[5] $p_1, p_2, p_3, \ldots$, and the logical connectives $\neg, \wedge, \vee$, listed in the preface on page ix.

We often use different labels for our variables (e.g., $a, b, c, \ldots, x, y, z, \ldots, p, q, r, \ldots$, etc.) as "metavariables" that stand for variables, and we define propositional formulas by structural induction: any variable p is a formula, and if α, β are formulas, then so are $\neg\alpha, (\alpha \wedge \beta)$, and $(\alpha \vee \beta)$. For example, $p, (p \vee q), (\neg(p \wedge q) \wedge (\neg p \vee \neg q))$. Recall also from the preface, that $\rightarrow$ and $\leftrightarrow$ are the implication and equivalence connectives, respectively.

Problem C.1. *Define propositional formulas with a context-free grammar. See section 4.6.3 for a refresher on context-free grammars.*

Lemma C.2. *Assign weights to all symbols as in figure C.1. The weight of any formula α is -1, but the weight of any proper initial segment is ≥ 0. Hence no proper initial segment of a formula is a formula.*

[5] Propositional variables are sometimes called *atoms*. A very thorough, and perhaps now considered a little bit old fashioned, discussion of "names" in logic (what is a "variable," what is a "constant," etc.), can be found in [Church (1996)], sections 01 and 02.

symbol	weight
$\neg$	0
$\wedge, \vee, ($	1
$), p$, for each variable p	-1

Fig. C.1 Assignments of "weights" to symbols.

Proof. By structural induction on the length of α. The basis case is: $w(p) = -1$, for any variable p. The induction step has three cases: $\neg\alpha$, $(\alpha \wedge \beta)$ and $(\alpha \vee \beta)$. This shows that any well-formed formula has weight -1. We now show that any proper initial segment has weight ≥ 0. In the basis case (a single variable p) there are no initial segments; in the induction step, suppose that the claim holds for α and β (that is, any initial segment of α, and any initial segment of β, has weight ≥ 0). Then the same holds for $\neg\alpha$, as any initial segment of $\neg\alpha$ contains $\neg$ (and $w(\neg) = 0$) and some (perhaps empty) initial segment of α. □

Problem C.3. *Finish the details of the proof of lemma C.2.*

Let $\alpha \overset{\text{syn}}{=} \alpha'$ emphasize that α and α' are equal as string of symbols, i.e., we have a syntactic identity, rather than a semantic identity.

Theorem C.4 (Unique Readability Theorem). *Suppose $\alpha, \beta, \alpha', \beta'$ are formulas, c, c' are binary connectives, and $(\alpha c \beta) \overset{\text{syn}}{=} (\alpha' c' \beta')$. Then $\alpha \overset{\text{syn}}{=} \alpha'$ and $\beta \overset{\text{syn}}{=} \beta'$ and $c \overset{\text{syn}}{=} c'$.*

Note that this theorem says that the grammar for generating formulas is unambiguous. Or, put another way, it says that there is only one candidate for the main connective, which means that the parse tree of any formula is unique. Recall that in problem 1.11 we compared infix, prefix, postfix notations; Boolean formulas are given in infix notation in the sense that the binary operators $(\wedge, \vee)$ are placed in between the operands, and yet it is unambiguous (whereas problem 1.11 says that we need two out of three representations, from among {infix,prefix,postfix}, to represent a tree unambiguously). The difference is that in the case of Boolean formulas we have parentheses to delimit subformulas.

Problem C.5. *Show that theorem C.4 is a consequence of lemma C.2. (Hint: define the weight of a formula to be the sum of the weights of all the symbols in it.)*

A *truth assignment* is a map τ : {variables} $\longrightarrow \{T, F\}$. Here $\{T, F\}$ represents "true" and "false," sometimes denoted 0,1, respectively. The truth assignment τ can be extended to assign either T of F to every formula as follows:

(1) $(\neg\alpha)^\tau = T$ iff $\alpha^\tau = F$
(2) $(\alpha \wedge \beta)^\tau = T$ iff $\alpha^\tau = T$ and $\beta^\tau = T$
(3) $(\alpha \vee \beta)^\tau = T$ iff $\alpha^\tau = T$ or $\beta^\tau = T$

The following are standard definitions: we say that the truth assignment τ *satisfies* the formula α if $\alpha^\tau = T$, and τ *satisfies* a set of formulas Φ if τ satisfies all $\alpha \in \Phi$. In turn, the set of formulas Φ is *satisfiable* if some τ satisfies it; otherwise, Φ is *unsatisfiable.* We say that α is a *logical consequence* of Φ, written $\Phi \models \alpha$, if τ satisfies α for every τ such that τ satisfies Φ. A formula α is *valid* if $\models \alpha$, i.e., $\alpha^\tau = T$ for all τ. A valid propositional formula is called a *tautology.* α and β are *equivalent* formulas (written $\alpha \iff \beta$) if $\alpha \models \beta$ and $\beta \models \alpha$. Note that '$\iff$' and '$\leftrightarrow$' have different meanings: one is a semantic assertion, and the other is a syntactic assertion. Yet, one holds if and only if the other holds.

For example, the following are tautologies: $p \vee \neg p, p \rightarrow p, \neg(p \wedge \neg p)$. An instance of logical consequence: $(p \wedge q) \models (p \vee q)$. Finally, an example of equivalence: $\neg(p \vee q) \iff (\neg p \wedge \neg q)$. This last statement is known as the "De Morgan Law."

Problem C.6. *Show that if $\Phi \models \alpha$ and $\Phi \cup \{\alpha\} \models \beta$, then $\Phi \models \beta$.*

Problem C.7. *Prove the following* Duality Theorem*: Let α' be the result of interchanging $\vee$ and $\wedge$ in α, and replacing p by $\neg p$ for each variable p. Then $\neg\alpha \iff \alpha'$.*

Problem C.8. *Prove the* Craig Interpolation Theorem: *Let α and β be any two propositional formulas. Let Var(α) be the set of variables that occur in α. Let S = Var(α) $\cap$ Var(β). Assume S is not empty. If $A \rightarrow B$ is valid, then there exists a formula C such that Var(C) $= S$, called an "interpolant" such that $A \rightarrow C$ and $C \rightarrow B$ are both valid.*

One way to establish that a formula α with n variables is a tautology is to verify that $\alpha^\tau = T$ for all 2^n truth assignments τ to the variables of α. A similar exhaustive method can be used to verify that $\Phi \models \alpha$ (if Φ is finite). Another way, is to use the notion of a formal proof; here we present the PK proof system, due to the German logician Gentzen (PK abbreviates "Propositional Kalkül").

In the propositional sequent calculus system PK, each line in a proof is a *sequent* of the form:

$$S = \alpha_1, \ldots, \alpha_k \rightarrow \beta_1, \ldots, \beta_l$$

where $\rightarrow$ is a new symbol, and $\alpha_1, \ldots, \alpha_k$ and $\beta_1, \ldots, \beta_l$ are sequences of formulas ($k, l \geq 0$) called *cedents* (*antecedent* and *succedent*, respectively).

A truth assignment τ *satisfies* the sequent S iff τ falsifies some α_i or τ satisfies some β_i, i.e., iff τ satisfies the formula:

$$\alpha_S = (\alpha_1 \wedge \cdots \wedge \alpha_k) \rightarrow (\beta_1 \vee \cdots \vee \beta_l)$$

If the antecedent is empty, $\rightarrow \alpha$ is equivalent to α, and if the succedent is empty, $\alpha \rightarrow$ is equivalent to $\neg\alpha$. If both antecedent and succedent are empty, then $\rightarrow$ is false (unsatisfiable).

We have the analogous definitions of validity and logical consequence for sequents. For example, the following are valid sequents: $\alpha \rightarrow \alpha$, $\rightarrow \alpha, \neg\alpha$, $\alpha \wedge \neg\alpha \rightarrow$.

A formal *proof* in PK is a finite rooted tree in which the nodes are labeled with sequents. The sequent at the root (bottom) is what is being proved: the *endsequent*. The sequents at the leaves (top) are *logical axioms*, and must be of the form $\alpha \rightarrow \alpha$, where α is a formula. Each sequent other than the logical axioms must follow from its parent sequent(s) by one of the rules of inference listed in figure C.2.

Problem C.9. *Give PK proofs for each of the following valid sequents:* $\neg p \vee \neg q \rightarrow \neg(p \vee q)$, $\neg(p \vee q) \rightarrow \neg p \wedge \neg q$, *and* $\neg p \wedge \neg q \rightarrow \neg(p \vee q)$, *as well as* $(p_1 \wedge (p_2 \wedge (p_3 \wedge p_4)))) \rightarrow ((((p_1 \wedge p_2) \wedge p_3) \wedge p_4)$.

Problem C.10. *Show that the contraction rules can be derived from the cut rule (with exchanges and weakenings).*

Problem C.11. *Suppose that we allowed $\leftrightarrow$ as a primitive connective, rather than one introduced by definition. Give the appropriate left and right introduction rules for $\leftrightarrow$.*

For each PK rule, the sequent on the bottom is a logical consequence of the sequent(s) on the top; call this the *Rule Soundness Principle.* For example, in the case of $\vee$-right it can be shown as follows: suppose that τ satisfies the top sequent; suppose now that it satisfies Γ. Then, since τ satisfies the top, it has to satisfy one of Δ, α or β. If it satisfies Δ we are done; if it satisfies one of α, β then it satisfies $\alpha \vee \beta$ and we are also done.

Problem C.12. *Check the Rule Soundness Principle: check that each rule is sound, i.e., the bottom of each rule is a logical consequence of the top.*

Weak structural rules

exchange-left: $\dfrac{\Gamma_1, \alpha, \beta, \Gamma_2 \to \Delta}{\Gamma_1, \beta, \alpha, \Gamma_2 \to \Delta}$ exchange-right: $\dfrac{\Gamma \to \Delta_1, \alpha, \beta, \Delta_2}{\Gamma \to \Delta_1, \beta, \alpha, \Delta_2}$

contraction-left: $\dfrac{\Gamma, \alpha, \alpha \to \Delta}{\Gamma, \alpha \to \Delta}$ contraction-right: $\dfrac{\Gamma \to \Delta, \alpha, \alpha}{\Gamma \to \Delta, \alpha}$

weakening-left: $\dfrac{\Gamma \to \Delta}{\alpha, \Gamma \to \Delta}$ weakening-right: $\dfrac{\Gamma \to \Delta}{\Gamma \to \Delta, \alpha}$

Cut rule

$$\frac{\Gamma \to \Delta, \alpha \quad \alpha, \Gamma \to \Delta}{\Gamma \to \Delta}$$

Rules for introducing connectives

$\neg$-left: $\dfrac{\Gamma \to \Delta, \alpha}{\neg\alpha, \Gamma \to \Delta}$ $\neg$-right: $\dfrac{\alpha, \Gamma \to \Delta}{\Gamma \to \Delta, \neg\alpha}$

$\wedge$-left: $\dfrac{\alpha, \beta, \Gamma \to \Delta}{(\alpha \wedge \beta), \Gamma \to \Delta}$ $\wedge$-right: $\dfrac{\Gamma \to \Delta, \alpha \quad \Gamma \to \Delta, \beta}{\Gamma \to \Delta, (\alpha \wedge \beta)}$

$\vee$-left: $\dfrac{\alpha, \Gamma \to \Delta \quad \beta, \Gamma \to \Delta}{(\alpha \vee \beta), \Gamma \to \Delta}$ $\vee$-right: $\dfrac{\Gamma \to \Delta, \alpha, \beta}{\Gamma \to \Delta, (\alpha \vee \beta)}$

Fig. C.2 PK rules. Note that Γ, Δ denote finite sequences of formulas.

Theorem C.13 (PK Soundness). *Each sequent provable in PK is valid.*

Proof. We show that the endsequent in every PK proof is valid, by induction on the number of sequents in the proof. For the basis case, the proof is a single line; an axiom $\alpha \to \alpha$, and it is obviously valid. For the induction step, one need only verify for each rule, if all top sequents are valid, then the bottom sequent is valid. This follows from the Rule Soundness Principle. □

The following is known as the *Inversion Principle*: for each PK rule, except weakening, if the bottom sequent is valid, then all top sequents are valid.

Problem C.14. *Inspect each rule, and prove the Inversion Principle, and give an example, with the weakening rule, for which this principle fails.*

Theorem C.15 (PK Completeness). *Every valid propositional sequent is provable in PK without using cut or contraction.*

Proof. We show that every valid sequent $\Gamma \to \Delta$ has a PK proof, by induction on the total number of connectives $\wedge, \vee, \neg$, occurring in $\Gamma \to \Delta$.

Basis case: zero connectives, so every formula in $\Gamma \to \Delta$ is a variable, and since it is valid, some variable p must be in both Γ and Δ. Hence $\Gamma \to \Delta$ can be derived from $p \to p$ by weakenings and exchanges.

Induction Step: suppose γ is not a variable, in Γ or Δ. Then it is of the form $\neg\alpha, (\alpha \wedge \beta), (\alpha \vee \beta)$. Then, $\Gamma \to \Delta$ can be derived by one of the connective introduction rules, using exchanges.

The top sequent(s) will have one fewer connective than $\Gamma \to \Delta$, and are valid by the Inversion Principle; hence they have PK proofs by the induction hypothesis. □

Problem C.16. *What are the five rules* not *used in the induction step in the above proof?*

Problem C.17. *Consider* PK$'$, *which is like* PK, *but where the axioms must be of the form $p \to p$, i.e., α must be a variable in the logical axioms. Is* PK$'$ *still complete?*

Problem C.18. *Suppose that $\{\to \beta_1, \ldots, \to \beta_n\} \models \Gamma \to \Delta$. Give a* PK *proof of $\Gamma \to \Delta$ where all the leaves are either logical axioms $\alpha \to \alpha$, or one of the non-logical axioms $\to \beta_i$. (Hint: your proof will require the use of the cut rule.) Now give a proof of the fact that given a finite Φ such that $\Phi \models \Gamma \to \Delta$, there exists a PK proof of $\Gamma \to \Delta$ where all the leaves are logical axioms or sequents in Φ. This shows that* PK *is also* Implicationally Complete.

C.2 First Order Logic

First Order Logic is also known as Predicate Calculus. We start by defining a *language* $\mathcal{L} = \{f_1, f_2, f_3, \ldots, R_1, R_2, R_3, \ldots\}$ to be a set of function and relation symbols. Each function and relation symbol has an associated *arity*, i.e., the number of arguments that it takes. *$\mathcal{L}$-terms* are defined by structural induction as follows: every variable is a term: $x, y, z, \ldots, a, b, c, \ldots$;

if f is an n-ary function symbol and $t_1, t_2, \dots, t_n$ are terms, then so is $ft_1t_2 \dots t_n$. A 0-ary function symbol is a constant (we use c and e as a metasymbols for constants). For example, if f is a binary (arity 2) function symbol and g is a unary (arity 1) function symbol, then $fgex, fxy, gfege$ are terms.

Problem C.19. *Show the Unique Readability Theorem for terms. See theorem C.4 for a refresher of unique readability in the propositional case.*

For example, the language of arithmetic, so called Peano Arithmetic, is given by $\mathcal{L}_A = [0, s, +, \cdot; =]$. We use infix notation (defined on page 3) instead of the formal prefix notation for $\mathcal{L}_A$ function symbols $\cdot, +$. That is, we write $(t_1 \cdot t_2)$ instead of $\cdot t_1 t_2$, and we write $(t_1 + t_2)$ instead of $+t_1t_2$. For example, the following are $\mathcal{L}_A$-terms: $sss0, ((x + sy) \cdot (ssz + s0))$. Note that we use infix notation with parentheses, since otherwise the notation would be ambiguous.

We construct $\mathcal{L}$-formulas as follows:

(1) $Rt_1t_2 \dots t_n$ is an *atomic formula*, R is an n-ary predicate symbol, $t_1, t_2, \dots, t_n$ are terms.
(2) If α, β are formulas, then so are $\neg\alpha, (\alpha \vee \beta), (\alpha \wedge \beta)$.
(3) If α is a formula, and x a variable, then $\forall x\alpha$ and $\exists x\alpha$ are also formulas.

For example, $(\neg\forall xPx \vee \exists x\neg Px)$, $(\forall x\neg Qxy \wedge \neg\forall zQfyz)$ are first order formulas.

Problem C.20. *Show that the set of $\mathcal{L}$-formulas can be given by a context-free grammar.*

We also use the infix notation with the equality predicate; that is, we write $r = s$ instead of $= rs$ and we write $r \neq s$ instead of $\neg = rs$.

An occurrence of x in α is *bound* if it is in a subformula of α of the form $\forall x\beta$ of $\exists x\beta$ (i.e., in the *scope* of a quantifier). Otherwise, the occurrence is *free*. For example, $\exists y(x = y + y)$: x is free, but y is bound. In $Px \wedge \forall xQx$ the variable x occurs both as free and bound. A term t or formula α are *close* if they contain no free variables. A closed formula is called a *sentence*.

We now present a way of assigning meaning to first order formulas: *Tarski semantics*; we are going to use the standard terminology and refer to Tarski semantics as the *basic semantic definitions* (BSD).

A *structure* (or *interpretation*) gives meaning to terms and formulas. An $\mathcal{L}$ structure $\mathcal{M}$ consists of:

(1) A nonempty set M, called the *universe of discourse.*
(2) For each n-ary f, $f^{\mathcal{M}} : M^n \longrightarrow M$.
(3) For each n-ary P, $P^{\mathcal{M}} \subseteq M^n$.

If $\mathcal{L}$ contains $=$, $=^{\mathcal{M}}$ must be the usual $=$. Thus equality is special—it must always be the true equality. On the other hand, $<^{\mathcal{M}}$ could be anything, not necessarily the order relation we are used to.

Every $\mathcal{L}$-sentence becomes either true or false when interpreted by an $\mathcal{L}$-structure $\mathcal{M}$. If a sentence α becomes true under $\mathcal{M}$, we say $\mathcal{M}$ *satisfies* α, or $\mathcal{M}$ is a *model* for α, and write $\mathcal{M} \vDash \alpha$.

If α has free variables, then they must get values from M (the universe of discourse), before α can get a truth value under $\mathcal{M}$. An *object assignment* σ for a structure $\mathcal{M}$ is a mapping from variables to the universe M. In this context, $t^{\mathcal{M}}[\sigma]$ is an element in M—given by the structure $\mathcal{M}$ and the object assignment σ. $\mathcal{M} \vDash \alpha[\sigma]$ means that $\mathcal{M}$ satisfies α when its free variables are assigned values by σ.

This has to be defined very carefully; we show how to compute $t^{\mathcal{M}}[\sigma]$ by structural induction:

(1) $x^{\mathcal{M}}[\sigma]$ is $\sigma(x)$
(2) $(ft_1t_2 \ldots t_n)^{\mathcal{M}}[\sigma]$ is $f^{\mathcal{M}}(t_1^{\mathcal{M}}[\sigma], t_2^{\mathcal{M}}[\sigma], \ldots, t_n^{\mathcal{M}}[\sigma])$

If x is a variable, and m is in the universe of discourse, i.e., $m \in M$, then $\sigma(m/x)$ is the same object assignment as σ, except that x is mapped to m. Now we present the definition of $\mathcal{M} \vDash \alpha[\sigma]$ by structural induction:

(1) $\mathcal{M} \vDash (Pt_1 \ldots t_n)[\sigma]$ iff $(t_1^{\mathcal{M}}[\sigma], \ldots, t_n^{\mathcal{M}}[\sigma]) \in P^{\mathcal{M}}$.
(2) $\mathcal{M} \vDash \neg\alpha[\sigma]$ iff $\mathcal{M} \nvDash \alpha[\sigma]$.
(3) $\mathcal{M} \vDash (\alpha \overset{(\vee)}{\wedge} \beta)[\sigma]$ iff $\mathcal{M} \vDash \alpha[\sigma]$ $\overset{\text{(or)}}{\text{and}}$ $\mathcal{M} \vDash \beta[\sigma]$.
(4) $\mathcal{M} \vDash (\overset{(\exists)}{\forall} x\alpha)[\sigma]$ iff $\mathcal{M} \vDash \alpha[\sigma(m/x)]$ for $\overset{\text{(some)}}{\text{all}}$ $m \in M$.

If t is closed, we write $t^{\mathcal{M}}$; if α is a sentence, we write $\mathcal{M} \vDash \alpha$.

For example, let $\mathcal{L} = [;R,=]$ (R binary predicate) and let $\mathcal{M}$ be an $\mathcal{L}$-structure with universe $\mathbb{N}$ and such that $(m,n) \in R^{\mathcal{M}}$ iff $m \leq n$. Then, $\mathcal{M} \vDash \exists x\forall yRxy$, but, $\mathcal{M} \nvDash \exists y\forall xRxy$.

The *standard structure* $\underline{\mathbb{N}}$ for the language $\mathcal{L}_A$ has universe $M = \mathbb{N}$, $s^{\underline{\mathbb{N}}}(n) = n+1$, and $0, +, \cdot, =$ get their usual meaning on the natural numbers. For example, $\underline{\mathbb{N}} \vDash \forall x\forall y\exists z(x+z = y \vee y+z = x)$, but $\underline{\mathbb{N}} \nvDash \forall x\exists y(y+y = x)$.

We say that a formula α is *satisfiable* iff $\mathcal{M} \vDash \alpha[\sigma]$ for some $\mathcal{M}$ & σ. Let Φ denote a set of formulas; then, $\mathcal{M} \vDash \Phi[\sigma]$ iff $\mathcal{M} \vDash \alpha[\sigma]$ for all $\alpha \in \Phi$.

$\Phi \models \alpha$ iff for all $\mathcal{M}$ & σ, if $\mathcal{M} \models \Phi[\sigma]$ then $\mathcal{M} \models \alpha[\sigma]$, i.e., α is a *logical consequence* of Φ. We say that a formula α is *valid*, and write $\models \alpha$, iff $\mathcal{M} \models \alpha[\sigma]$ for all $\mathcal{M}$ & σ. We say that α and β are *logically equivalent*, and write $\alpha \iff \beta$, iff for all $\mathcal{M}$ & σ, ($\mathcal{M} \models \alpha[\sigma]$ iff $\mathcal{M} \models \beta[\sigma]$).

Note that $\models$ is a symbol of the "meta language" (English), as opposed to $\wedge, \vee, \exists, \ldots$ which are symbols of first order logic. Also, if Φ is just one formula, i.e., $\Phi = \{\beta\}$, then we write $\beta \models \alpha$ in place of $\{\beta\} \models \alpha$.

Problem C.21. *Show that* $(\forall x\alpha \vee \forall x\beta) \models \forall x(\alpha \vee \beta)$, *for all formulas* α *and* β.

Problem C.22. *Is it the case that* $\forall x(\alpha \vee \beta) \models (\forall x\alpha \vee \forall x\beta)$?

Suppose that t, u are terms. Then:

$t(u/x)$ result of replacing all occurrences of x in t by u

$\alpha(u/x)$ result of replacing all *free* occurrences of x in α by u

Semantically, $(u(t/x))^{\mathcal{M}}[\sigma] = u^{\mathcal{M}}[\sigma(m/x)]$ where $m = t^{\mathcal{M}}[\sigma]$.

For example, let $\mathcal{M}$ be $\underline{\mathbb{N}}$ (the standard structure) for $\mathcal{L}_A$. Suppose $\sigma(x) = 5$ and $\sigma(y) = 7$. Let:

u be the term $x + y$

t be the term $ss0$

Then:

$$u(t/x) \text{ is } ss0 + y \text{ and so } (u(t/x))^{\underline{\mathbb{N}}}[\sigma] = 2 + 7 = 9$$

On the other hand, $m = t^{\underline{\mathbb{N}}} = 2$, so $u^{\underline{\mathbb{N}}}[\sigma(m/x)] = 2 + 7 = 9$.

Problem C.23. *Prove* $(u(t/x))^{\mathcal{M}}[\sigma] = u^{\mathcal{M}}[\sigma(m/x)]$ *where* $m = t^{\mathcal{M}}[\sigma]$, *using structural induction on* u.

Problem C.24. *Does the result in problem C.23 apply to formulas* α? *That is, is it true that* $\mathcal{M} \models \alpha(t/x)[\sigma]$ *iff* $\mathcal{M} \models \alpha[\sigma(m/x)]$, *where* $m = t^{\mathcal{M}}[\sigma]$?

For example, suppose α is $\forall y\neg(x = y + y)$. This says "x is odd". But $\alpha(x + y/x)$ is $\forall y\neg(x + y = y + y)$ which is always false, regardless of the value of $\sigma(x)$. The problem is that y in the term $x + y$ got "caught" by the quantifier $\forall y$.

A term t is *freely substitutable for* x *in* α iff no free occurrence of x in α is in a subformula of α of the form $\forall y\beta$ or $\exists y\beta$, where y occurs in t.

Theorem C.25 (Substitution Theorem). *If t is freely substitutable for x in α then for all structures $\mathcal{M}$ and all object assignments σ, it is the case that $\mathcal{M} \vDash \alpha(t/x)[\sigma]$ iff $\mathcal{M} \vDash \alpha[\sigma(m/x)]$, where $m = t^{\mathcal{M}}[\sigma]$.*

Problem C.26. *Prove the Substitution Theorem. (Hint. Use structural induction on α and the BSDs.)*

If a term t is not freely substitutable for x in α, it is because some variable y in t gets caught by a quantifier $\forall y$ or $\exists y$ in α. One way to fix this is simply rename the bound variable y in α to some new variable z. This renaming does not change the meaning of α.

Let $a, b, c, \ldots$ denote free variables and let $x, y, z, \ldots$ to denote bound variables. A first order formula α is called a *proper formula* if it satisfies the restriction that it has no free occurrence of any "bound" variable and no bound occurrence of any "free" variable. Similarly a *proper term* has no "bound" variable. Notice that a subformula of a proper formula is not necessarily proper, and a proper formula may contain terms which are not proper.

The sequent system LK is an extension of the propositional system PK where now all formulas in the sequent $\alpha_1, \ldots, \alpha_k \to \beta_1, \ldots, \beta_l$ must be proper formulas. The system LK is PK together with the four rules for introducing quantifiers given in figure C.3.

$$\forall \text{ introduction:} \quad \frac{\alpha(t), \Gamma \to \Delta}{\forall x \alpha(x), \Gamma \to \Delta} \qquad \frac{\Gamma \to \Delta, \alpha(b)}{\Gamma \to \Delta, \forall x \alpha(x)}$$

$$\exists \text{ introduction:} \quad \frac{\alpha(b), \Gamma \to \Delta}{\exists x \alpha(x), \Gamma \to \Delta} \qquad \frac{\Gamma \to \Delta, \alpha(t)}{\Gamma \to \Delta, \exists x \alpha(x)}$$

Fig. C.3 Extending PK to LK.

There are some restrictions in the use of the rules given in figure C.3. First, t is a proper term, and $\alpha(t)$ (respectively, $\alpha(b)$) is the result of substituting t (respectively, b) for all free occurrences of x in $\alpha(x)$. Note that t, b can be freely substituted for x in $\alpha(x)$ because $\forall x \alpha(x), \exists x \alpha(x)$ are proper formulas. The free variable b must not occur in the conclusion in $\forall$ right and $\exists$ left.

Problem C.27. *Show that the four new rules are sound.*

Problem C.28. *Give a specific example of a sequent* $\Gamma \to \Delta, \alpha(b)$ *which is valid, but the bottom sequent* $\Gamma \to \Delta, \forall x\alpha(x)$ *is not valid, because the restriction on* b *is violated (*b *occurs in* Γ *or* Δ *or* $\forall x\alpha(x)$*). Do the same for* $\exists$ *left.*

An LK proof of a valid first order sequent can be obtained using the same method as in the propositional case. Write the goal sequent at the bottom, and move up using the introduction rules in reverse. If there is a choice about which quantifier to remove next, choose $\forall$ right or $\exists$ left (working backward), since these rules carry a restriction.

C.3 Peano Arithmetic

Recall the language of arithmetic, $\mathcal{L}_A = [0, s, +, \cdot; =]$. The axioms for PA are the following

P1 $\forall x(sx \neq 0)$
P2 $\forall x \forall y(sx = sy \to x = y)$
P3 $\forall x(x + 0 = x)$
P4 $\forall x \forall y(x + sy = s(x + y))$
P5 $\forall x(x \cdot 0 = 0)$
P6 $\forall x \forall y(x \cdot sy = x \cdot y + x)$

plus the *Induction Scheme:*

$$\forall y_1 \dots \forall y_k[(\alpha(0) \land \forall x(\alpha(x) \to \alpha(sx))) \to \forall x\alpha(x)] \tag{C.1}$$

where α is any $\mathcal{L}_A$-formula, and (C.1) is a sentence. Note that this is the formal definition of induction given in section 1.1.

We also have a scheme of equality axioms.

E1 $\forall x(x = x)$
E2 $\forall x \forall y(x = y \to y = x)$
E3 $\forall x \forall y \forall z((x = y \land y = z) \to x = z)$
E4 $\forall x_1 \dots \forall x_n \forall y_1 \dots \forall y_n(x_1 = y_1 \land \dots \land x_n = y_n) \to fx_1 \dots x_n = fy_1 \dots y_n$
E5 $\forall x_1 \dots \forall x_n \forall y_1 \dots \forall y_n(x_1 = y_1 \land \dots \land x_n = y_n) \to Px_1 \dots x_n \to Py_1 \dots y_n$

where E4 and E5 hold for all n-ary function and predicate symbols. In $\mathcal{L}_A$, which is our language of interest, s is unary, $+, \cdot$ are binary, and $=$ is binary.

Let LK-PA be LK where the leaves are allowed to be P1-6 and E1-5, besides the usual axioms $\alpha \rightarrow \alpha$. For example, $\rightarrow \forall x(x = x)$ would be a valid leaf.

Problem C.29. *Show that LK-PA proves that all nonzero elements have predecessor.*

Problem C.30. *Show that LK-PA proves the following: the associative and commutative law of addition, the associative and commutative laws of multiplication and that multiplication distributes over addition. Specify carefully which axioms you are using.*

C.4 Answers to selected problems

Problem C.1. Let the grammar G_{prop} have the alphabet $\{p, 1, \wedge, \vee, \neg\, (,)\}$, and of the set of rules given by

$$S \longrightarrow pX|\neg S|(S \wedge S)|(S \vee S)$$
$$X \longrightarrow 1|X1$$

The variables are $\{p1, p11, p111, p1111, \ldots\}$, i.e., they are encoded in unary notation.

Problem C.3. By the induction hypothesis, $w(\alpha) = w(\beta) = 1$, so $w(\neg\alpha) = 0 + (-1) = -1$ and since the left and right parentheses balance each other out, in the sense that $w((t)) = w(() + w(t) + w()) = 1 + w(t) + (-1) = w(t)$, the result quickly follows for $(\alpha \wedge \beta)$ and $(\alpha \vee \beta)$. To show that any propor initial segment of $(\alpha \circ \beta)$ (where $\circ \in \{\wedge, \vee\}$) has weight ≥ 0, we write it as follows:

$$(\alpha \circ \beta) \stackrel{\text{syn}}{=} (\alpha_1\alpha_2 \ldots \alpha_m \circ \beta_1\beta_2 \ldots \beta_n)$$

where α_i and β_j are the symbols of α and β, respectively. Several cases naturally present themselves: if the initial segment consists only of (, then its weight is 1. If the initial segment ends in the α_i's, but does not end at α_m, then by induction it has weight ≥ 1. If it ends exactly at α_m, then by induction it has weight 0. If it ends at $\circ$, then it has weight 1. Similarly, we deal with the initial segment ending in the middle of the β_j's, at β_n, and at the last parenthesis).

Problem C.6. Suppose that we have $\Phi \models \alpha$ and $\Phi \cup \{\alpha\} \models \beta$. And suppose that τ is a truth assignment that satisfies Φ. Then, by the first assumption it must satisfy α, and so τ satisfies $\Phi \cup \{\alpha\}$, and hence by the second assumption it must satisfy β.

Problem C.7. By structural induction on α. Clearly, if α is just a variable p, then α' is $\neg p$, and $\neg\alpha \iff \alpha'$. The induction step follows directly from De Morgan Laws.

Problem C.8. Let the variables of α be $\alpha(\bar{x}, \bar{y})$ and the variables of β be $\beta(\bar{y}, \bar{z})$. The notation $\bar{x}$ denotes a set of Boolean variables; using this convention, the set $S = \mathrm{Var}(\alpha) \cap \mathrm{Var}(\beta) = \{\bar{y}\}$. We define the Boolean function f as follows:

$$f(\bar{y}) = \begin{cases} 1 & \text{if } \exists\bar{x} \text{ such that } \alpha(\bar{x}, \bar{y}) = 1 \\ 0 & \text{otherwise} \end{cases}$$

We are abusing notation slightly here, by mixing Boolean functions and Boolean formulas; $\bar{y}$ is working over-time: it is both an argument to f and a truth assignment to α. But the meaning is clear. Let $C_f(\bar{y})$ be the Boolean formula associated with f; it can be obtained, for example, by conjunctive normal form. The C_f is our formula: suppose that $\tau \vDash \alpha$; then τ clearly satisfies C_f (by its definition). If $\tau \vDash C$, then there must be an $\bar{x}$ such that $\alpha(\bar{x}, \tau)$ is true, and hence $\beta(\tau, \bar{z})$ is true by the original assumption.

Note that we could have defined f dually with β; how?

Problem C.21. We prove this using BSDs: Let $\mathcal{M}$ be any structure, and σ any object assignment. Suppose $\mathcal{M} \vDash (\forall x\alpha \vee \forall x\beta)[\sigma]$. Then, $\mathcal{M} \vDash \forall x\alpha[\sigma]$ or $\mathcal{M} \vDash \forall x\beta[\sigma]$.

Case (1): $\mathcal{M} \vDash \forall x\alpha[\sigma]$. Then, $\mathcal{M} \vDash \alpha[\sigma(m/x)]$ for all $m \in M$. Then, $\mathcal{M} \vDash (\alpha \vee \beta)[\sigma(m/x)]$ for all $m \in M$. So, $\mathcal{M} \vDash \forall x(\alpha \vee \beta)[\sigma]$.

Case (2): $\mathcal{M} \vDash \forall x\beta[\sigma]$; same idea as above.

Therefore, $\mathcal{M} \vDash \forall x(\alpha \vee \beta)[\sigma]$. By the definition of logical consequence, $(\forall x\alpha \vee \forall x\beta) \vDash \forall x(\alpha \vee \beta)$

Problem C.22. No, not necessarily. We use the def of logical consequence to prove this. To prove that the RHS is *not* a logical consequence of the LHS, we must exhibit a model $\mathcal{M}$, an object assignment σ and formulas α, β such that: $\mathcal{M} \vDash \forall x(\alpha \vee \beta)[\sigma]$, but $\mathcal{M} \nvDash (\forall x\alpha \vee \forall x\beta)[\sigma]$.

Let α and β be Px and Qx, respectively (P, Q unary predicates). Now define $\mathcal{M}$ and σ. Since the formulas are sentences, no need to define σ. $\mathcal{M}$: let the universe of discourse be $M = \mathbb{N}$. We still need to give meaning in $\mathcal{M}$ to P, Q. Let $P^{\mathcal{M}} = \{0, 2, 4, \ldots\}$, and $Q^{\mathcal{M}} = \{1, 3, 5, \ldots\}$. Then: $\mathcal{M} \vDash \forall x(Px \vee Qx)$ (because every number is even or odd).

But, $\mathcal{M} \nvDash (\forall xPx \vee \forall xQx)$ (because it is not true that either all numbers are even or all numbers are odd).

Problem C.24. For example, suppose α is $\forall y \neg(x = y + y)$. This says "$x$ is odd". But $\alpha(x + y/x)$ is $\forall y \neg(x + y = y + y)$ which is always false, regardless of the value of $\sigma(x)$. The problem is that y in the term $x + y$ got "caught" by the quantifier $\forall y$.

Problem C.29. Let $\alpha(x)$ be $(x = 0 \vee \exists y(x = sy))$. We outline the proof informally, but the proof can of course be formalized in LK-PA. Basis case: $x = 0$, and LK-PK proves $\alpha(0)$ easily:

$$\cfrac{\cfrac{\cfrac{\cfrac{\to \forall x(x=x)}{\to 0=0, \forall x(x=x)}\ \text{weak \& exch} \qquad \cfrac{0=0 \to 0=0}{\forall x(x=x) \to 0=0}\ \forall\text{-left}}{\to 0=0}\ \text{Cut}}{\to 0=0, \exists y(0=sy)}\ \text{weak}}{\to 0=0 \vee \exists y(0=sy)}\ \vee\text{-right}$$

Induction Step: Show that LK-PA proves $\forall x(\alpha(x) \to \alpha(sx))$, i.e., we must give an LK-PA proof of the sequent:

$$\to \forall x(\neg(x = 0 \vee \exists y(x = sy)) \vee (sx = 0 \vee \exists y(sx = sy)))$$

This is not difficult, and it is left to the reader. From the formulas $\alpha(0)$ and $\forall x(\alpha(x) \to \alpha(sx))$, and using the axiom:

$$\to (\alpha(0) \wedge \forall x(\alpha(x) \to \alpha(sx))) \to \forall x \alpha(x)$$

we can now conclude (in just a few steps): $\to \forall x \alpha(x)$ which is what we wanted to prove. Thus, LK-PA proves $\forall x \alpha(x)$.

C.5 Notes

There are many excellent introductions to logic; for example, [Buss (1998)] and [Bell and Machover (1977)]. This section follows the logic lectures given by Stephen Cook at the University of Toronto.

Problem C.7 is of course an instance of the general Boolean "Duality Principle." A proof-theoretic version of this principle is given, for example, as theorem 3.4 in [Mendelson (1970)], where the *dual* of a proposition concerning a Boolean algebra B is the proposition obtained by substituting $\curlyvee$ for $\curlywedge$ and $\curlywedge$ for $\curlyvee$ (see page 171 where we defined these symbols). We also substitute 0 for 1 and 1 for 0. Then, if a proposition is derivable from the usual axioms of Boolean algebra, so is its dual.

Bibliography

Agrawal, M., Kayal, N. and Saxena, N. (2004). Primes is in P, *Annals of Mathematics* **160**, 2, pp. 781–793.

Alford, W. R., Granville, A. and Pomerance, C. (1994). There are infinitely many Carmichael numbers, *Annals of Mathematics* **139**, 3, pp. 703–722.

Allan Borodin, C. R., Morten N. Nielsen (2003). (incremental) priority algorithms, *Algorithmica* **37**, 4, pp. 295–326.

Alperin, J. L. and Bell, R. B. (1995). *Groups and Representations* (Springer).

Arrow, K. (1951). *Social Choice and Individual Values* (J. Wiley).

Bell, J. and Machover, M. (1977). *A course in mathematical logic* (North-Holland).

Berkowitz, S. J. (1984). On computing the determinant in small parallel time using a small number of processors, *Information Processing Letters* **18**, 3, pp. 147–150.

Borodin, A. and El-Yaniv, R. (1998). *Online Computation and Competitive Analysis* (Cambridge University Press).

Buss, S. R. (1998). An introduction to proof theory, in S. R. Buss (ed.), *Handbook of Proof Theory* (North Holland), pp. 1–78.

Cenzer, D. and Remmel, J. B. (2001). Proof-theoretic strength of the stable marriage theorem and other problems, *Reverse Mathematics* , pp. 67–103.

Church, A. (1996). *Introduction to Mathematical Logic* (Princeton University Press).

Cormen, T. H., Leiserson, C. E., Rivest, R. L. and Stein, C. (2009). *Introduction to Algorithms* (McGraw-Hill Book Company), third Edition.

Delfs, H. and Knebl, H. (2007). *Introduction to Cryptography* (Springer).

Dijkstra, E. W. (1989). On the cruelty of really teaching computing science, *Communications of the ACM* **32**, 12.

Dorrigiv, R. and López-Ortiz, A. (2009). On developing new models, with paging as a case study, *ACM SIGACT News* **40**, 4.

Downey, A. (2008). *Think Python: How to Think Like a Computer Scientist* (Green Tea Press).

Dummit, D. S. and Foote, R. M. (1991). *Abstract Algebra* (Prentice Hall).

Engel, A. (1998). *Problem-Solving Strategies* (Springer).

Gale, D. and Shapley, L. S. (1962). College admissions and the stability of marriage, *American Mathematical Monthly* **69**, pp. 9–14.

Halmos, P. R. (1960). *Naive Set Theory* (Springer-Verlag).

Halmos, P. R. (1995). *Linear algebra problem book* (The mathematical association of America).

Hardy, G. H. and Wright, E. M. (1980). *An Introduction to the Theory of Numbers*, 5th edn. (Oxford University Press).

Harel, D. (1987). *Algorithmics: The Spirit of Computing* (The Addison-Wesley Publishing Company), ISBN 0-201-19240-3.

Hoffman, P. (1998). *The Man Who Loved Only Numbers: The Story of Paul Erdős and the Search for Mathematical Truth* (Hyperion).

Hoffstein, J., Pipher, J. and Silverman, J. H. (2008). *An Introduction to Mathematical Cryptography* (Springer).

Karp, R. M. and Rabin, M. O. (1987). Efficient randomized pattern-matching algorithms. *IBM Journal of Research and Development* **31**, 2, pp. 249–260.

Kleinberg, J. and Tardos, É. (2006). *Algorithm Design* (Pearson Education).

Knuth, D. E. (1997). *The Art of Computer Programming*, Vol. 1, Fundamental Algorithms, 3rd edn. (Addison Wesley).

Kozen, D. (2006). *Theory of Computation* (Springer).

Manna, Z. (1974). *Mathematical Theory of Computation* (McGraw-Hill Computer Science Series).

Mendelson, E. (1970). *Boolean algebra and switching circuits* (McGraw Hill).

Michael Aleknovich, J. B.-O. R. I. A. M. T. P., Allan Borodin (2005). Toward a model of backtracking and dynamic programming, .

Papadimitriou, C. H. (1994). *Computational Complexity* (Addison-Wesley).

Papadimitriou, C. H. and Steiglitz, K. (1998). *Combinatorial Optimization: Algorithms and Complexity* (Dover).

Press, W. H., Vetterling, W. T., Teukolsky, S. A. and Flannery, B. P. (2007). *Numerical Recipes: The Art of Scientifc Computing*, 3rd edn. (Cambridge University Press).

Reingold, O. (2005). Undirected st-connectivity in log-space, in *STOC'05: Proceedings of the thirty-seventh annual ACM symposium on Theory of computing*, pp. 376–385.

Rosen, K. H. (2007). *Discrete Mathematics and Its Applications*, 6th edn. (McGraw Hill).

Shustek, L. (2009). Interview, *Communications of the ACM* **52**, 3, pp. 38–41.

Singh, S. (1999). *The Code Bok: The evolution of secrecy, from Mary, Queen of Scots, to Quantum Cryptography* (Doubleday).

Sipser, M. (2006). *Introduction to the Theory of Computation* (Thompson), second Edition.

Solovay, R. and Strassen, V. (1977). A fast monte-carlo test for primality, *SIAM Journal of Computing* **6**, pp. 84–86.

Soltys, M. (2002). Berkowitz's algorithm and clow sequences, *Electronic Journal of Linear Algebra* **9**, pp. 42–54.

Soltys, M. (2009). *An Introduction to Computational Complexity* (Jagiellonian University Press), to appear.

van Vliet, H. (2000). *Software Engineering: Principles and Practice*, 2nd edn. (Wiley).

Velleman, D. J. (2006). *How To Prove It*, 2nd edn. (Cambridge University Press).

von zur Gathen, J. and Gerhard, J. (1999). *Modern computer algebra* (Cambridge University Press).

Zhai, Y. (2010). *Pairwise comparisons based non-numerical ranking*, Ph.D. thesis, McMaster University.

Zingaro, D. (2008). *Invariants: A generative approach to programming* (College Publications).

Index